Belarus

From Soviet Rule to Nuclear Catastrophe

David R. Marples

Professor of History
University of Alberta

MACMILLAN

First published 1996 by
MACMILLAN PRESS LTD
Houndmills, Basingstoke, Hampshire RG21 6XS
and London
Companies and representatives
throughout the world

ISBN 0–333–62631–1 hardcover
ISBN 0–333–62632–X paperback

A catalogue record for this book is available
from the British Library.

10   9   8   7   6   5   4   3   2   1
05 04 03 02 01 00 99 98 97 96

Printed in Great Britain by
Antony Rowe Ltd
Chippenham, Wiltshire

To my wife, Lan Chan-Marples

Contents

List of Plates

List of Tables

Acknowledgments

This study is based on interviews, conducted in the Republic of Belarus over the course of seven visits during the period 1992–5, with doctors, scientists, academicians and journalists; on official documents, journals, articles and newspaper sources from Belarus, 1986–94; on visits to the contaminated zone and interviews with families therein; and on periods of research in the National Library of Belarus in Minsk. It is also founded on research conducted during a Summer Research fellowship at the Center for Russian and East European Studies, University of Illinois at Urbana-Champaign in 1993; and (during a term as a professor at the Harvard University Summer School in 1994) at the Widener Library, Harvard University. It also includes information provided by two scientific conferences held under the auspices of the Belarusian Charitable Fund for the Children of Chernobyl in April 1992 and April 1994. My attendance at one of these meetings was financed by the University of Alberta's Central Research Fund. Colleagues from Belarus have sent me a great variety of scientific and academic papers over the past three years. Last but by no means least has been the remarkable clippings service provided by the Office of Public Information in Minsk, which has enabled the compiling of information from the most remote regions of the republic.

My work has benefitted from the financial support of the Professional Partnership Program, Association of Universities and Colleges of Canada, Ottawa; and two Support for the Advancement of Scholarship awards from the Faculty of Arts, University of Alberta which enabled the completion of the project in 1994–5. One visit to Belarus was financed with the assistance of the office of the Associate Vice-President (Research) of this same university. This is my fifth book with Macmillan and the association with its director, Tim Farmiloe, and a series of editors (currently Gráinne Twomey) has been a long and fruitful one.

My debts to friends and colleagues are many. In particular I would like to thank the following: in Belarus, Gennadiy Grushevoy, Iryna Grushevaya, E.P. Kanaplya, E.P. Demidchik, Yourie Pankratz, Sergey Laptev, Y. Stulov, K. Radyuk, Vladimir Glod, D.D. Kozikis, Rita Kozikis, V. Tyomkin, T. Tyomkina, N.A. Gres', Ludmilla Chistyakova, Dmitriy Chesnov, V.S. Kazakov, Vladimir Polishchuk, A.N. Stozharov, Iryna Ushakevich, and G.A. Nesvetailov; in the United States, Andrew Bond, Lisa Moskowitz, Joe Arciuch, Jan Zaprudnik and Jurij Dobczansky; and in Canada, Ernest McCoy, Doris Bradbury, Robert Busch, David F. Duke, and Allan Tupper. A special thanks goes to Yuri Novikov, a native of Minsk now resident in Alberta, who was my research assistant in Canada for part of 1994; and to David J. Hall, chairman of the Department of History and Classics at the University of Alberta, for his constant encouragement of my research over the past four years. My chief debt, however, is to my research assistant in Minsk for this entire study, Lyuba Pervushina. Without her unflagging efforts to arrange the various interviews and visits to the contaminated zone and other regions of Belarus, secure permission for me to visit archives and libraries, uncover numerous sources and many other tasks, this study could not have been completed.

Finally, I would like to thank my family for once again having the fortitude to endure a marathon period of research and study that necessitated so many visits abroad: my wife, Lan, to whom this book is dedicated; and my sons Carlton and Keelan.

Edmonton, Canada
May 1995

NOTE ON TRANSLITERATION

I have adhered throughout to Belarusian spelling of place names in the text, with the single exception of Minsk, which is an international city, and is not widely known by its Belarusian name of Miensk. I have also used as a guideline the excellent map issued by the Committee for Land Surveying with the Belarusian Council of Ministers in 1992. The map perhaps takes some liberties with historic names, hence one has Brest rather than Beras'tse. Concerning human subjects, if the person is clearly of Russian origin, then the Russian version of the name is retained. The transliteration system is that of the Library of Congress, though I have modified this slightly, preferring the use of the ending -ya to -ia; and the ending of names with -iy, as opposed to -ii. Belarusian linguists will not, I hope, find it offensive that this form of transliteration renders their language more similar to Russian than might otherwise be the case. In footnotes and references I have used the original spelling of the author or editor, hence if Belarusian authors wrote articles or books in Russian, the Russian form of their name is cited.

Map of Belarus (Courtesy of the *Belarusian Review*)

Introduction

In April 1992 I was invited to visit Minsk for the first time, at the behest of Dr Yourie Pankratz, a man I had met twice in the United States in April 1991, when we were both speakers at conferences on successive days in Washington, D.C. and Chicago. Both conferences had focused at least in part on the fifth anniversary of the Chernobyl disaster, a subject to which I had devoted two previous books. During his speeches in the two US cities, Dr Pankratz stressed that too little attention had been paid to his republic, Belarus, which had received the largest proportion of the radioactive fallout. Rather, largely because of the presence of a very active Ukrainian diaspora in North America, most of our focus had been directed toward Ukraine.

Upon arrival in Belarus, to attend another major international conference on Chernobyl in Minsk, the subject came up again during several meetings. Why were so few Western academics interested in this small republic? Scientists and medical specialists were certainly present; many from the Western countries and Japan had begun to visit Belarus on a regular basis. During the conference, and once I was recognized by those present after the delivery of a presentation, the Belarusians among the attendees – scientists, academicians and journalists – appeared almost to be lining up to provide me with their materials. The message was always the same: we need attention; we are suffering from the effects of a disaster and yet we are being ignored by the outside world.

Yet these were tumultuous times and Chernobyl was but one aspect of many attracting attention. Belarus was in its eighth month of independence. Less than four months had passed since the USSR had been dissolved, a result of a meeting in this republic between the leaders of the three Slavic states: Russia, Ukraine, and Belarus. Frankly, the political change was of more interest to me than the future results of Chernobyl, which I considered that I had covered adequately in previous studies. Hitherto I had spent most of my time conducting research on Ukraine. Unconsciously, therefore, I was constantly comparing life in the two republics. Ukraine was easier to fathom, particularly the western parts. Ukrainian intellectuals worked for the future of their new country. Even in Canada several hundred Ukrainian Canadians had departed for Ukraine either on a temporary or permanent basis: to start new businesses; to assist the government; to develop a legal reform program; to assist with new agricultural techniques. I could detect no such foreign presence in Minsk.

The Belarusians would tell jokes about Ukrainians, almost all based on what was to them an incomprehensible nationalism.

In the middle of the April 1992 conference, a small delegation, including myself, was suddenly granted an interview with the chairman of the parliament, Stanislau Shushkevich – often referred to erroneously in the Western media as the 'president' of Belarus. He was, nonetheless, the leading political figure in the republic at that time and, together with the presidents of Russia and Ukraine, had been responsible for the foundation of the so-called Commonwealth of Independent States (CIS) that effectively forced the dissolution of the Soviet Union in December 1991. The parliament buildings are located in the central square (formerly Lenin Square, now Independence Square), a vast draughty concourse, with a small cobblestone area with a grassy park adjacent to it in the midst of two wide lanes of traffic. A large statue of Lenin glowers angrily over the square from its position outside the parliament building, looking toward the offices of the government. Further along, but still part of – and overshadowed by – this huge complex is the red Polish Catholic cathedral (being renovated at the time) and the offices of the Minsk City Council.

There was almost no security to be seen. I wandered with a companion through the parliament building at will until we found the designated meeting room. Among our delegation was a Canadian of Belarusian descent, Ivanka Survilla, a proudly nationalistic woman and a gifted linguist; while our party was led by an opposition parliamentary deputy and the chairman of the charitable fund 'For the Children of Chernobyl' (discussed in detail in this volume), Gennadiy Grushevoy, the organizer of our conference. The stocky balding figure of Shushkevich swept into the room, businesslike, and, speaking in Russian, demanded questions. At once, Ms Survilla accosted him. In fluent Belarusian she demanded to know whether he had forgotten his native language. Shushkevich, a professor of physics by training, immediately switched to Belarusian, though he was visibly taken aback. That a member of the diaspora had sought to remind the leader of the republic of his 'national' duties was not merely symbolic; it reflected two disparate views of the new post-Soviet world and Belarusian society in particular.

The other aspect of the interview that interested me was Shushkevich's utter lack of tolerance for Grushevoy. Shushkevich simply refused to recognize his presence there, though he was courteous to the rest of our delegation, which included Americans, Japanese, Germans and three Belarusian interpreters. The formal politeness to all visitors that would have been taken for granted in the Canadian parliament was, however, absent. Just as inexplicably, from my perspective, the two politicians who had freely traded insults that day would be reconciled prior to the 1994 presidential campaign. Shushkevich was altogether a curiosity because he did not seem to be either on the side of the Communist majority in parliament or the opposition of the Belarusian Popular Front. Yet he was manifestly popular. Whom did he represent?

By the end of the conference, I had resolved to study further the situation in Belarus. As a historian, however, I was dissatisfied with the idea of a straightforward examination of the results of Chernobyl. To carry out such a study seemed to me to neglect the very foundations of the problem, which could be encapsulated in a simple, single question: what is Belarus? I wanted to ascertain whether there were Belarusians who were devoutly nationalistic; whether they approached their lives as Belarusians first and foremost; what their attitude might be to the Russians and other groups of the former Soviet Union, perhaps especially the Lithuanians with whom they were linked historically. I had found no ready explanations during the April 1992 visit.

Returning in October 1992, as the guest of Dr. Yuriy Stulov, then the Vice Rector of the Institute of Foreign Languages (now the Minsk State Linguistic University), I was able to have some profitable discussions with local academicians and others about how they perceived the new state. Many were nonplussed by the questions I asked. They responded that the question of national identity was immaterial to them, as was the language they used in the street and at home, which was invariably Russian, with a few Belarusian phrases thrown in for good measure. Some were stridently pro-Russian. There was evidently far more interest in events in Moscow than those in Minsk. I attributed this attitude to the volatile nature of the relationship between the Russian president and his parliament.

During one visit I was taken to two sites of historical importance. One was Khatyn, the memorial for Belarusian victims of the Great Patriotic War, located about 50 kilometers from Minsk. As one approaches the memorial, a huge statue is encountered, of an old grieving man holding a dead child. Only three members of that village had survived (two others had been wounded and left for dead). A bell tolled constantly. The memorial to the victims of the war was well maintained. It pointed out to visitors the extent of the losses in World War II, when one in four residents of the Belarusian SSR had died. I wondered immediately why the site of Khatyn – a name eerily similar to the site of the massacred Polish officers at Katyn, which is only two hours' drive away – should have been chosen. Was it to confuse visitors over the two sites? I was informed by several locals that the name of the memorial site was not significant; that it was merely a coincidence that the two names were so alike.

The second memorial was a more recent one; the mass gravesite at Kurapaty, which is on the outskirts of Minsk in a broad forest belt. Though both sites date to the same period, the late 1930s and early 1940s, the contrast could not have been more profound. A furious battle was being waged with the Communist authorities to gain official recognition of the Kurapaty site. Many prominent Communists simply refused to acknowledge that the NKVD was responsible for the executions at Kurapaty, despite the findings of the archeologist Zyanon Paz'nyak. That Paz'nyak, who rediscovered the site in 1988, had become the leader of the Belarusian Popular Front also made the findings unpalatable to certain sectors of society: the government; the former partisans and NKVD officials; and war veterans in general. Something was wrong here. Mass gravesites were being discovered throughout the former USSR; why was it that only in Belarus was there such unwillingness to investigate the historical facts?

Kurapaty preceded directly the start of the 'Great Patriotic War,' an event that might be described, along with Chernobyl, as one of the formative events in the history of twentieth-century Belarus. One cannot travel anywhere in the republic without being reminded of the war, and the sacrifice of the Soviet people. Belarus was occupied by the Germans for longer than any other part of the USSR. It also suffered proportionally the highest losses of any Soviet republic and became the center of the partisan movement. I had no intention in this volume of examining these issues in great depth since they merit a separate study. I did want, nonetheless, to study them as key factors in the development of the current state and how the population views that state and its future within it. The initial question was whether there was even a future at all.

In the summer of 1993, I decided to return to the republic on instinct, travelling from Warsaw by train without a visa. At the border the train stopped at a small platform and swarms of KGB men boarded. They were polite, but at Brest, two returned and I was escorted off the train to obtain my visa. I noticed that their uniforms now had Belarus epaulettes. One asked me why I had decided to visit Belarus. When I informed him that I was a historian, he was shocked. 'Why are you interested in this republic?' he wanted to know, 'There is no history here.' The statement remained with me. In Minsk I perused every bookstore I could find for historical works. There were hardly any. The store clerk was surprised when I asked for some.

Amid the drabness of the city of Minsk, I still felt that there was a mystery to be uncovered. What had happened to these people that they had lost interest in their own history? Why had they lost their language?

Was it yet another manifestation of the processes of Russification? Yet if the latter was the case, why was there so little anger toward Moscow? I had discovered little to suggest that the Russians were regarded as anything other than the closest of friends. There was also an atmosphere of resignation, the abandonment of hope on various issues, whether they be the economic situation (in 1992–3 Belarus was noticeably less affected by inflation and unemployment than Ukraine) or the impact of the nuclear disaster at Chernobyl. Most families complained about their state of health, or of friends and relatives who had contracted an illness. Chernobyl, it was evident, had become if nothing else a huge psychological burden upon the population. It was regarded as a 'sword of Damocles' waiting to strike down the future, coming generations. Even Alyaksandr Lukashenka, the first president, who soon sought almost dictatorial powers at the expense of the parliament, felt obliged to pay attention to Chernobyl and in particular the plight of the children affected by the disaster.

There were thus a number of intriguing issues: the language question; the purges of the Stalin period; the war; postwar development; Chernobyl; and the emergence of an independent state. Were these contemporary issues somehow connected in the history of twentieth-century Belarus? Was there some link between them all? The premise of this book is that there is a clear connection between state development in the Soviet period, and the events since 1986. The population was an integral part of a Soviet state. It could not for the most part be separated from the interests, needs or care of that state. The state was not there to be judged, and Belarusians were not in a position to make such judgments. Only a few cries in the wilderness, from fanatics, 'bourgeois nationalists', and others who acted against the interests of the majority of the people suggested that there were any flaws in this uniformity. The state had created the Byelorusian SSR; and through the unity of the Soviet peoples it had defeated and destroyed the most dangerous threat to the twentieth-century world: Nazi Germany.

I could imagine – though it was never any more than idle musing – what life must have been like for these citizens 30 and 40 years earlier, as they made their life plans, attended work, socialized, often in the belief that their lives were unchanging, secure, and that the future held bright hopes. In 1961 Khrushchev had outlined a future utopia with a new party program at the 22nd Congress of the Central Committee of the Communist Party of the Soviet Union. Shortly, he prognosticated, the villages would experience all the benefits of the towns. I could thus conceive of past life before everything came to such a dramatic end. In the National Library I came across journals that depicted Gorbachev and Yeltsin as monsters who had betrayed Soviet ideals and ultimately destroyed Khrushchev's dream. On the streets in 1993, a new newspaper had emerged called *Nash kompas*. Its masthead depicted Marx, Engels, Lenin and Stalin. Perhaps, I thought, in Belarus it is the Stalin era that is most revered since that was the time of greatest achievement. The circulation of *Nash kompas* was only 10 000, but its sentiments were not significantly different from those expressed in the popular evening newspaper, *Vecherniy Minsk*. Even though the Communist Party after 1991 was a shadow of its former self, a substantial number of citizens had become nostalgic about the Soviet period. All these matters intrigued me and played some role in the production of this volume.

In terms of research work, my debt to those historians of Belarus who have trodden this route before me must not be underestimated. There are precious few of them, but they have made a significant contribution to our knowledge. They include Jan Zaprudnik, the former head of research for the Belarusian program at Radio Liberty, with whom I have had the opportunity to talk frequently in recent years, particularly at academic conferences and briefings. Zaprudnik's book, *Belarus: At a Crossroads in History* (the full citation is provided in the Bibliography) brought Belarus back into the mainstream of Soviet and post-Soviet studies. Nicholas Vakar, Ivan Lubachko and Steven Guthier can be regarded as pioneers of Belarusian history in the West. All three embarked on major studies at a time when the area was even less popular than it is today. More recently at Radio Liberty in Munich, prior to its metamorphosis into the Open Media Research Institute, Kathleen Mihalisko conducted a *tour-de-force* of personal research on the contemporary Belarusian scene that this author always found intelligent and useful. The profound comments of Thomas Bird of the City University of New York on the state of the Church in Belarus – offered at three conferences that I attended in the period 1992–4 – have not been utilized in this volume, but they have assisted my background knowledge of this complex subject.

The above scholars have paved the way for academics and graduate students of the future. Despite their efforts, the road ahead seems very much an uncovered one, and there are likely to be many pitfalls. In the future, one suspects, it may become 'unfashionable' to devote one's time to geographical regions or countries of the world; rather themes and trends within a variety of areas may be compared and related. In the Belarusian case, such a development would be unfortunate in that despite centuries of history, there are as yet very few studies that have encompassed this republic. There are gaps and there are, as local historians of the republic describe them, many 'blank spots' that need to be filled. It is hoped that this study is a contribution to that end.

This book therefore is intended not merely as an analysis of the effects of Chernobyl, though that theme is a dominant one. It seeks to put the contemporary problems of Belarus into historical perspective. It argues that modern problems have old roots; that today's nihilistic philosophy and what to a Westerner appears as appalling pessimism about the future owes much to the way the area and the state were developed in the twentieth century. Without wishing to antagonize Belarusian patriots, it might be postulated that the Soviets helped both to create and destroy Belarus and that the republic in its present state lacks a rational basis for existence. In such an atmosphere a tragedy such as Chernobyl takes on new meaning. It augments and supplements existing problems. It must be stated at the outset: the Belarusian population does not for the most part treat Chernobyl in a rational manner. But this was hardly to be expected. There is no question that we are dealing with a tragedy of unprecedented dimensions.

1 Soviet Rule: Repression and Urbanization

The objective of this chapter is to examine the evolution of Belarus[1] toward independence, and outline its distinctive characteristics as a republic of the former Soviet Union. The republic of Belarus, it is posited, is a state that has been developed with some unique features both historically and as a result of economic and political measures imposed by its Soviet rulers from the 1920s to 1991. Arguably these features have thus far impeded its evolution into a non-Communist and democratic state oriented toward market reforms, while at the same time the dominance and stability of its ruling elite – partly as a result of that elite's manipulation of state structures – may also have prevented its degeneration into the civil conflict that has occurred in most other post-Soviet regimes.

The lack of fundamental economic and political reform has been evident in Belarus after four years of independence. The question most frequently asked by visitors is: Why has there been so little change in Belarus since the Soviet period? Why does a visit to Minsk seem in many respects a reacquaintance with the Soviet past, even to the extent that statues of Lenin and Cheka founder Feliks Dzerzhinsky (who was born in what is today Belarusian territory) adorn the central square and main street respectively? Is there something inherently different about the Belarusians that has caused them to lag so markedly in the development of national consciousness and state development on a national basis? Has this lack of national consciousness had an appreciable effect upon the state's ability to deal with the repercussions of a nuclear catastrophe? Should one take special note of the fact that during the presidential election campaign of 1994, the eventual landslide winner, Alyaksandr Lukashenka, emphasized the fact that he had been the only parliamentary deputy opposed to the December 1991 Belavezhski (Brest Oblast) agreement between the Russian and Ukrainian presidents and Belarusian parliamentary speaker Stanislau Shushkevich dissolving the Soviet Union; or that a policy of reunion with Russia was one of the most popular electoral slogans?

Since 1989, it should be noted, there has been a tendency in central and eastern Europe to forgo the structures of the Soviet era; to inculcate new thinking in economic policy, and in countries such as Poland and Hungary to embark boldly on a new course of capitalism. Nations that were held together through 'marriages of convenience' (Czechoslovakia and Yugoslavia are the two most notable examples) have split asunder, partly because of emergent national sentiment. The merits or problems of these developments need not be debated here. Our emphasis is on the contrast between these nations and the Republic of Belarus. Other former Soviet nations have also placed national above economic interests; in republics such as Moldova, two hostile states exist alongside each other, distinguished primarily by ethnic composition. In the central region of the Caucasus, Chechnya bitterly resisted the Russian army in 1994–5, in an effort to maintain an existence as an independent state – independence was declared in 1991 – after 130 years of Russian rule. These are complex issues and here they are generalized somewhat. They do nevertheless draw one again to the main point: none of these developments has occurred in Belarus.

The election of Lukashenka drew forth several warnings from observers that the new president was likely to take the state back to the darkest Communist era. Lukashenka, a former state farm manager born in the Vitsebsk Oblast, whose career was developed in Mahileu, was perhaps a compromise candidate. His position as chairman of the parliamentary Commission on Corruption had provided him with powerful leverage for a political campaign, not least because through the commission he succeeded in discrediting his considerably older rivals: the former Speaker of the parliament, Stanislau Shushkevich; and the former prime minister, Vyachaslau Kebich, also a Communist by background. After a quiet few months in office (also referred to by the parliamentary opposition as a 100-day grace period), Lukashenka began to lash out at the reformers in his Cabinet, and particularly at the market-oriented chief of the National Bank, Stanislau Bahdankevich. By the fall of 1994, the economic picture in the republic became unequivocally gloomy, with the currency in a free fall against the US dollar, and with an estimated 50 percent of the population living below the poverty line.[2] The key question is whether this new republic has the will or the means to continue its progress as an independent entity and, if so, whether this entity can break away from the Communist past and conformist attitude *vis-à-vis* its neighbor Russia.

GEOGRAPHY

Belarus is a landlocked republic of approximately 10 353 000. Its total area is 207 600 square kilometers wedged between Poland to the west, Russia to the north, Ukraine to the south, and Lithuania and Latvia to the northwest. It consists of six *oblasts* (provinces): Brest, Vitsebsk, Homel', Hrodna, Minsk, and Mahileu, with a total of 118 *raions* (districts). The average density of population is 49.7 persons per square kilometer, and about 53 percent of the population is female. There is no official religion in the republic, though the Belarusian Orthodox Church predominates, with a substantial minority of Roman Catholics, and smaller groups of Uniates and others. Its population percentages (the subject will be dealt with in detail below) in 1989 consisted of Belarusians 77.9, Russians 13.2, Poles 4.1, Ukrainians 2.9, and Jews 1.1.[3]

Belarus's most distinctive features are its humid climate, its flatness and the prevalence of the forest belt, dominated by pine trees, which encompasses about one-third of the total territory of the modern republic. The forests are widely dispersed across the country rather than centered in a major location, hence though it is possible to see trees no matter where one travels, one cannot discern a large-scale contiguous forest.[4] Belarus is a land of rivers and lakes. The major river is the Dnyapro (Dnieper), which flows almost the length of the republic, with its important tributaries the Pripyats' in the south and the Byarezina in the central region. Its second major river is the Dzvina in the north; while the Neman links western Belarus with Lithuania. In the far west, the Buh crosses into Belarusian territory, passing through the city of Brest. The podzolic soil which prevails in two-thirds of the country is not ideal for agriculture and up to 25 percent of the territory is covered with swamps and peat bogs.[5] Nevertheless, Belarus accounted for an impressive 20 percent of the USSR's flax fiber production (on humid soils), and 15 percent of its potatoes (grown on dry, sandy soils).[6] In terms of mineral deposits, Belarus is a relatively poor country and in the past was dependent for fuel largely on supplies of local peat,[7] though it has some largely undeveloped reserves of hard coal, natural gas and oil.

HISTORIOGRAPHY

Belarus is a republic that to date has received relatively little attention from scholars, and is particularly lacking major studies by historians. Some efforts were made to address this problem in the 1950s–1970s by scholars writing in English,[8] though for some twenty years there did not appear to my knowledge a single full-length scholarly English-language monograph devoted exclusively to Belarus until the appearance of the pioneering volume by Jan Zaprudnik in 1993.[9] In the Soviet period in Belarus, there appeared the standard histories devoted to the 'successes' of the Soviet state and the amicable relations between the Belarusian and Russian republics. One such work intended for a foreign audience (it was published in Paris under the auspices of UNESCO) is worth citing as typical of the era and illustrative of the superficial nature of such works.

The book, entitled *Cultural Policy in the Byelorussian Soviet Socialist Republic*, is devoted to cultural policy in the BSSR and was published in 1979. It focuses on the main developments in the history of the republic in the Soviet period. Following the formation of the BSSR on 1 January 1919, it states, the BSSR resolved to unite with the other Soviet republics within the USSR in December 1922. Illiteracy was eradicated after several years of Soviet rule. The Soviet era is divided into three stages. In the 'first period' from the October Revolution to the 1930s the main task of the authorities was to introduce a 'socialist attitude' among the population and to develop the ideological base of the working class. The peasants were to be provided with a 'Soviet' outlook on life and were to develop a collective spirit. The new culture was to be 'national in form and socialist in content.'[10]

Stage 2, according to this official history, encompassed the period from the 1930s to the 1950s, during which the building of socialism was completed and the new national leaders fully versed in the basic tenets of Marxism-Leninism. The victory over fascism occurred during 1941–5 and the war left 2.23 million residents of Belarus dead, ruined over 10 000 factories, and destroyed 80 percent of urban dwellings and 9200 villages. Stage 3 covered the final stage of the cultural revolution in Belarus and prepared the way for the transition from socialism to full communism. The authors noted that industrial production was raised dramatically in the years of the 9th Five-Year Plan (1971–5) with significant rises in national income and real per capita income. Through its industrial development, Belarus took its place as a valuable and prosperous Soviet republic.[11]

In this study and others, especially those of the Brezhnev era (1964–82), the portrayal of the development of Belarus by Soviet scholars is unremarkable when compared to that of any other Soviet republic. There is no explanation why Belarus was developed as an industrial base and selected to be one of the manufacturing centers of the Soviet Union. Little or no attention is given to specifically national issues, such as the development of the native language within the general cultural revolution and the importance of Belarusian culture and historical traditions. Admittedly far more sophisticated studies appeared than the above and some are worthy of close inspection and even citation. Yet the above remained typical prior to the late 1980s. Historians of the Soviet period were not willing to explore in depth the events of 1917 or the historical legacies of Belarus that did not originate with Russia or reveal a close relationship between the Belarusians and Russians.

The Great Patriotic War is one of the few events that received a separate or individual treatment for Belarus in Soviet works. The importance of the war to the former regime is symbolized by the four-storey Museum of the Great Patriotic War in Minsk, which is impressive in size and holdings by any standards; and the war memorial at Khatyn, some 40 miles from Minsk, which remains one of the better preserved memorials in the former Soviet Union. The memorial commemorates not only the destruction of the village which bore this name, but also the more than 2200 other settlements destroyed during the German–Soviet conflict and the extremely high Belarusian casualties (an estimated 25 percent of the prewar population). Official interpretations of the war, then, may have served to enhance loyalty to the Soviet regime and to establish a pivotal role for Belarus as one of the republics to have suffered most from that event.[12] Subsequently, the Soviet regime ensured that the sacrifices of that time were never forgotten and equated the actions of the main partisan groups closely with Belarus as a republic. In turn, former partisans played a decisive role in leadership politics in the BSSR in the early postwar years.[13]

The best example of this phenomenon, and one that has received attention from Soviet and post-Soviet historians in Belarus, is the life and career of the former partisan Piotr Mironovich Masherau, one of the genuinely popular figures to emerge from the Communist Party leadership. This is not the place to attempt a full-scale assessment of Masherau, a figure who has perhaps been idealized against the background of Brezhnev-style *apparatchiki*.[14] Suffice it to say that as with the partisans, the charismatic leader, who wore traditional costumes and was known to speak Belarusian on public occasions, gave further credibility to the Soviet regime in the BSSR. When Masherau was killed in a car accident in 1980, there was reportedly genuine mourning in the republic.[15] The Stalin purges and the (often unknown) mass executions of the 1930s and 1940s were thus offset by the wartime experiences and the long-term party leadership of a populist, Masherau.

The idealization of Masherau by historians is, however, misleading. Despite his promotion of Belarusian interests, in the final analysis, Masherau did not deviate significantly from central policies laid down in Moscow. In September 1978, for example, at a Plenum of the Central Committee of the Communist Party of Belarus (CPB), held three months after Brezhnev had visited Minsk to award the city the Order of Lenin and to present Masherau with the title 'Hero of Socialist Labor' on his sixtieth birthday, Masherau delivered the main report. Having outlined in typical

Brezhnev-style fashion the successes in Belarusian industry, he then focused on what he termed the problem of 'local patriotism' allegedly catalyzed by economic achievements. Masherau warned Belarusians that they must not view their work in isolation from the single Soviet economic complex and noted that, 'Our duty ... is to pay tribute above all to the great Russian people ... .'[16] Ultimately, then, Masherau posed as a Soviet patriot and a loyal supporter of the Brezhnev administration. Belarus remained a stable segment of the USSR even under his leadership. Belarusian historians of this period have thus provided a distorted portrait in order to promote endearment of the Soviet regime, and more recently the memory of that regime, among readers.

The result of these developments was that in the summer of 1991, the newly declared independent state lacked a historiographical foundation outside the Soviet context. The history of Belarus as an individual entity was not perceived as something that within the republic merited study in its own right. The Soviet regime, and first and foremost the Russian people had, according to the prevailing dogma, established the BSSR, helped it to survive invasion and occupation through the formation of partisan detachments, eliminated illiteracy, and built up in the republic an important industrial base. The concomitant concealment of a national history was enhanced by the purges of the 1930s, which effectively decapitated the Belarusian intelligentsia, and by the war years which in a very real sense saw the destruction of the national elite of the republic.

There have been some attempts to alleviate this deficiency in historical tradition. In 1992, the noted work by the historian Usievalad Ihnatouski (Vsevolod Ignatovskiy) of 1926, covering the history of the Belarus from earliest times to the start of the Soviet period, was republished in the Belarusian language.[17] The author was one of the founders and proponents of the BSSR, and in 1929 he had been appointed president of the Belarusian Academy of Sciences. However, his work was banned in 1929 during Stalin's campaign against National Communists, and Ignatouski was arrested and purged on the grounds that he had been a member of a (fictitious) anti-Soviet organization. He committed suicide in February 1931.[18] The work's appearance was thus an effort to rehabilitate one of Stalin's victims in Belarus. In addition to the republication of Ihnatouski's magisterial work, in 1988 a four-volume *History of the Working Class of Belorussia* was issued, while reportedly in this same year a *History of the Peasantry of Belorussia* was under preparation.[19]

In the spring of 1993, the Belarusian newspaper *Zvyazda* published a remarkable series of contributions from leading republican historians under the general title '100 Questions and Answers on the History of

Belarus,' which was subsequently published in a Russian version in the mass circulation newspaper *Narodnaya hazeta* (26 and 27 May, 1 June 1993).[20] Some of the questions raised were so basic as to suggest an audience that was largely ignorant of its own history. Examples include: 'When was Christianity established in Belarus?' 'Was there Mongol–Tatar enslavement in Belarus?' 'Is it true that the Lithuanians conquered Belarus?' The historians, however, went to some lengths to answer a question that well illustrates the official Soviet line for previous decades: 'Was the Belarusian People's Republic a puppet state?', explaining that the formation of the BPR represented a genuine and legitimate stage in the development of a national state.[21]

In this same year, two historians published a book on Belarusian history consisting once again of a question and answer format and explored new ground by outlining some of the problems within the partisan movement, and the imposition of Stalinist purges within its ranks. It was pointed out, for example, that on Belarusian territory two noted partisan leaders, V.I. Nichiporovich and V.S. Pyzhikov, were arrested and one was executed, on the orders of Stalin's NKVD chief, Lavrentiy Beria.[22] Generally, however, such works are only today beginning to broach the most sensitive subjects in twentieth-century Belarusian history. The 'myths' of the Soviet period may remain intact as long as prominent figures associated with the old regime manage to stay in control of the ruling apparatus, or if others such as the current president adhere to a naive nostalgia for the Soviet regime. By the same token, it is still a difficult task to uncover impartial and objective treatments of a Belarusian national tradition within the republic.

The history of Belarus has been left largely to the diaspora of North America and Europe. Yet Belarusian emigres are considerably fewer in number than their Ukrainian or Baltic counterparts. Indeed the number of specifically Belarusian institutions devoted to the study of the country and its history are very few: a Belarusian Academy of Arts and Sciences operates in New York; the well-edited *Belarusian Review* is published in Los Angeles. Belarusian studies are not widely taught at western universities, however. Even in the post-Soviet era, and with the emergence of an independent state, such studies must be considered marginal at best.[23]

FEATURES OF STATE DEVELOPMENT

We will highlight what are perceived to be key issues within the process of state development: first, the urbanization and industrialization of Soviet

Belarus; second, Stalinism and its principal legacies (to preserve chronological composure, the discussion of this question is included within the previous section); third (in Chapter 2), the question of language, and whether the formation and maintenance of a national state requires also the preservation of a state language, or, perhaps, two state languages. Do Belarusians of various backgrounds consider the present state as their homeland; and are they committed to the preservation of independence? The answers to such questions assist in providing a framework for the future state and a response to such pertinent queries as whether it can survive as an independent entity and, if so, what might its place be within central or eastern Europe. Must Belarus turn to the east, that is, Russia, for guidance and economic stability, or are there alternatives to the west and north?

The purpose here is not to provide an outline of Belarusian history in the twentieth century *per se*, but rather to stress features of development that appear to be distinctive to this republic, and which have thereby had a significant impact on the contemporary situation. Belarus did have a tradition of statehood prior to the twentieth century, but not a durable one or one that could be used as a precedent for the modern nation. Nor did it possess an indigenous and growing workforce in urban centers prior to Soviet rule. In this respect, it differed from European regions of Russia, which saw a remarkable growth of industry in the period from the 1880s to the turn of the twentieth century. Belarusians remained predominantly rural, living in peasant communities where illiteracy rates were high. Two of the leading cities in the territories of Soviet and post-Soviet Belarus – Vitsebsk and Minsk – had Yiddish-speaking majorities at the turn of the century according to the 1897 census; while a third, Vilna (Wilno, Vilnius), had a 40 percent Yiddish-speaking population, 31 percent Polish-speaking and 20 percent Russian-speaking. In all cities in today's territory of the republic, only Mahileu possessed a significant proportion of Belarusian speakers (29.8 percent) at the turn of the century.[24] (One can perceive a similar situation in the ethnic Ukrainian territories of the Austrian Empire, where in cities such as L'viv (L'vov, L'wow, Lemberg), Poles and Jews made up a majority of the city population and Ukrainians had barely penetrated city life.) Table 1.1 provides the results of the 1897 census in urban areas in the five Belarusian *gubernias* of the Russian Empire (the Northwest Province).

In the late nineteenth century, there were, nevertheless, some significant economic developments, particularly in industry and railroad construction. In industry, the areas that make up today's republic saw the establishment of light industry, including distilleries. In 1862, the first railroad was com-

Table 1.1   *National composition of the city population in Belarusian regions in 1897 (in percentages of total population)*

Gubernia	Belarusians	Russians	Poles	Jews
Vilna	7.6	18.3	26.6	43.0
Vitsebsk	13.3	22.1	8.1	52.0
Hrodna	9.6	15.8	13.3	61.2
Minsk	11.8	19.5	7.2	58.8
Mahileu	30.0	13.0	2.9	52.4

Source: Kasperovich, 1985, p. 31.

pleted as part of a line connecting St Petersburg and Warsaw, and subsequently further lines were constructed to link Moscow with Brest. There were reportedly some 8000 industrial and railroad workers in Belarusian territories by the turn of the century, but in contrast to St Petersburg with its massive industrial enterprises, the Belarusian factories were small affairs, and the province became a supplier of raw materials and farm products. According to a Soviet source, per capita industrial output in the Belarusian regions was less than 50 percent of the average for the Russian empire by the outbreak of the First World War; and in the sphere of machine building, ten times lower than the national average.[25] Zaprudnik observes that unemployment and poverty in this period led to large-scale emigration of Belarusians to the United States.[26] According to another source, Belarusians also migrated eastward into Siberia and central Russia. It is estimated that in the twenty years prior to the 1917 Revolutions, 520 000 peasants moved to the eastern regions of Russia and about 100 000 emigrated abroad.[27]

Nonetheless, it would be misleading to paint a completely bleak picture of industrial development in the period 1900–13. Evidence suggests that, on the contrary, significant progress was made in industries in the territories of Belarus, both in terms of output and numbers of workers. For example, the labor force involved in heavy industry rose from 31 000 in 1900 to almost 55 000 on the outbreak of the First World War. Industrial output increased in this same period by 129 percent. The number of small factories and enterprises employing less than 16 workers rose slightly, but at a much slower rate than the very large factories of over 1000 workers. The number of workers in the latter enterprises thus rose from 8 percent to

12.4 percent in 1900–13. By 1913 the total urban workforce in Belarus was about 155 000.[28] The development of this part of the Russian Empire was undeniable; Belarusian involvement in this process, however, was minimal.

For the most part, the Belarusian sector of the population remained occupied in agricultural pursuits. Based on figures from the 1897 census, Steven L. Guthier notes that 98 percent of native Belarusians lived in the countryside or in settlements with populations numbering less than 2000. In communities with more than 2000 people, only 16.1 percent of residents were ethnic Belarusians.[29] As Table 1.1 indicates, the Jewish population predominated in the cities of Belarus. For the Jews, this situation was not necessarily by choice. The first migration of Jews into what is today Belarus occurred in the eighth century, and these migrants were assimilated into the local population. A second wave followed from West European territories, mainly in the late sixteenth century.[30] In the late eighteenth century, the government of Imperial Russia established a 'Pale of Settlement' in which the Jews were obligated to reside, and limited to the 15 western *gubernias* of the Russian Empire. Between 1825 and 1861, the number of urban residents in Belarus increased to 316 000, or by 2.1 times. Part of the reason for this growth was the resettlement of the Jewish population from villages to large towns so that already by 1850, the Jews made up more than 50 percent of city traders.[31] The Jews of the Belarusian territories, like their counterparts elsewhere, were not permitted to acquire land outside cities or to live beyond city borders. Consequently, they constituted most of the traders and artisans of Belarusian towns.

After the Jews, the largest ethnic segment in the cities was comprised of the Russians, particularly in the eastern regions bordering Russia proper. There had been several waves of Russian migration into the Belarusian *gubernias* and concomitant Russification of these areas. In the late eighteenth and early nineteenth centuries, the Tsarist government sent Russian landowners and officials to settle lands in the western province. In addition some Russian peasants arrived independently, buying up lands from landowners and officials, or from those who had left to work in the cities.[32] As for the Poles, their numbers allegedly contained large numbers of Belarusians who had converted to Catholicism and become polonized.[33]

The foundation of the first independent Belarusian state emerged as a result of the Russian Revolutions of 1917 yet initially at least appeared to contravene the goals of Lenin and the Bolsheviks by the establishment of a so-called 'bourgeois regime.' Though this period saw a revival of national state traditions, the number of nationally conscious Belarusians remained small. The Bolsheviks, on the other hand, regarded Belarus as a buffer region between Soviet Russia and a hostile Poland (with which Russia was at war until 1920). An autonomous Belarusian state within Soviet Russia – and particularly one that was ruled by non-Bolsheviks – was unacceptable to the Bolshevik leaders and perhaps also to some of the large number of non-Belarusian urban residents in the ethnically Belarusian territories. At this time, there was no indigenous urban Belarusian elite in existence. When Belarus did declare itself a national republic on 9 March 1918, and subsequently an independent state on 25 March it was already occupied by the Germans.[34] Although the new state issued several decrees it soon became evident that the Germans were not willing to tolerate the new government.

With the departure of the Germans following their defeat on the Western Front in World War I, the territories were subject to a change of policy in Petrograd, resulting eventually in the formation, announced in Smolensk, of a Belarusian Soviet Socialist Republic (BSSR) on 1 January 1919.[35] Four days later this government was moved to Minsk, which became the capital of the BSSR. The new regime was ratified on 1 February. Within a month, this regime was forcibly superseded by a new Lithuanian-Belarusian Republic (Litbel) announced in Vilna. While Minsk and Vilna districts were included in this new entity, the Russian Federation swallowed up the eastern districts of the short-lived BSSR.[36]

The establishment of a Soviet regime in Belarus was not finalized, however, until the Polish–Soviet war ended in July 1920. As a result of this war, the Russians were able to retain the eastern part of the ethnic Belarusian territories, in addition to the city of Minsk. In August 1920, the BSSR was once again proclaimed, though drastically reduced in size to six *raions* in Minsk Oblast. A formal treaty was signed with Soviet Russia on 16 January 1921, and on 30 December 1922, the BSSR – in its truncated form – formally joined the Union of Soviet Socialist Republics. In doing so, it had conceivably gained more than the republics of the Transcaucasus which were not permitted to form individual Soviet republics, but most matters of significance were controlled by Moscow, as a result of the 1921 Treaty. Soviet works stressed that the state was formed democratically by the Belarusian workers and peasants.[37] Such a state, formed and controlled by Moscow, could hardly suffice in the forthcoming era of National Communism.

Ironically, however, it was the Bolsheviks in the 1920s who solidified and expanded the Belarusian SSR to encompass Belarusian ethnic territories that had been initially included into Soviet Russia,[38] established its capital in Minsk, and deliberately nurtured and encouraged the development of national culture in the period of the New Economic Policy (NEP).

This is not to say that in the 1920s a majority of Belarusian citizens would necessarily have supported the principle of a Belarusian state (though some did); merely that by creating this republic and expanding its initial territory, the Soviet authorities helped to establish and deliberately promote a feeling of distinctiveness among Belarusians. There was also even by the mid-1920s an inflow of native Belarusians into cities, and particularly to the capital, Minsk. Yet by the time of the 1926 census, this development can hardly be called of major significance. As Jan Zaprudnik notes, 91 percent of Belarusians were still rural dwellers and Belarusians continued to constitute a minority within the urban population.[39]

The early Soviet period (the 1920s and 1930s) saw two concomitant and significant developments in the formation of a national republic. First, Belarus was developed both industrially and culturally, and a more substantial beginning was made in the process of creating a national culture and national elite. Second, and seemingly the antithesis of the first development, was a movement toward strict centralization of command at the center (Moscow), with severe repressions conducted in Belarus as elsewhere commencing in 1930 among all sectors of society. One can take this a step further by suggesting that the non-Russian Slavic republics were regarded as essential to the maintenance of Soviet power and nurturing of Communist ideology in the urban regions. Repressions in Belarus might take the form of purges, public executions, deportations or dekulakization during the collectivization of agriculture, and were in some respects more severe than in most other territories of the Soviet Union. Let us look at these developments in chronological perspective.

There is no question that the 1920s were years of cultural revival for Belarus. In July 1924, an official program was drawn up to develop the republic culturally and to make the Belarusian language a language of daily usage in every facet of life (see below). Lubachko points out that by 1 February 1927, the 'Belarusification' of the government, with the tolerance of the central authorities in Moscow, had made significant progress and encompassed 100 percent of the Central Executive Committee, 100 percent of the Council of People's Commissars, 100 percent of the Commissariat of Education, 50 percent of the Commissariat of Agriculture, 30 percent of the Commissariat of Internal Affairs, and 30–50 percent of all other commissariats. Hence Belarusians were being permitted to play a dominant role in the leadership organs of their republic. Lubachko labels this development – with the concomitant growth of Belarusian schools – as a 'Golden Age' of Belarusian culture.[40]

The 1920s also saw significant demographic change in the BSSR. In brief, there was a large migration of Belarusians from the villages into the cities. In addition, there was a policy of moving residents of other republics, especially the Russian Federation, into the cities of the BSSR, ostensibly because of the shortage of labor. By the time of the 1926 census, the Jewish numerical domination of cities had virtually ended. Though Jews still constituted a plurality in urban centers, the proportion of other nationalities had increased so markedly as to indicate a definite trend rather than a statistical aberration. Jews now constituted just over 40 percent of the total urban population; Belarusians 39 percent; and Russians 15.6 percent. However, only 8 percent of all Belarusians actually lived in urban centers compared to 34 percent of Russians and 83 percent of Jews. Thus the cities of the early BSSR were cosmopolitan in composition with a variety of languages spoken, though even in the 1920s, a relatively tolerant period of Soviet rule, Russian tended to predominate.

Industrialization was stepped up rapidly in the first two Stalinist Five-Year Plans (1928–32 and 1933–37) when over 1700 industrial plants were constructed in the BSSR. By 1940, it is estimated, output of heavy industry exceeded that of the year 1913 by 14 times.[41] Still, the figures need to be put into perspective. The prewar industrial development in the republic was not as advanced as that in Russia or Ukraine, which were made the focus of industrialization in the European part of the USSR.[42] The most dramatic social transformation awaited the Khrushchev and Brezhnev eras.

THE IMPACT OF STALINISM

Kurapaty and Vileyka

The question of the impact of Stalinism on Belarus is still very much under review and little consensus has been reached within the republic. Stalinism took the form of mass executions of the population, particularly in the period 1937–41; the incorporation of Western Belarus and the very thorough depolonization of this territory; and a more systematic eradication of Belarusian intellectuals and cultural figures, in addition to those linked in any way to the short-lived Belarusian People's Republic of 1918. The executions were so extensive that they have been labelled as genocide by one writer because of the way in which they served to eliminate the intelligentsia of a nation.[43] The difficulties involved in unravelling the various events is typified by the discovery of and investigation into the massacres at Kurapaty, a forest just to the north of the city of Minsk.[44] Despite seemingly overwhelming evidence to indicate NKVD involvement in the shootings in this area, the Belarusian authorities have bowed to pressure from those who have claimed that the German occupation authorities were responsible. Kurapaty is perhaps the supreme example of the continuing 'blank spots' in Belarusian history. Despite overwhelming evidence to the contrary, the authorities thus far have managed to distort the truth about this massacre. In so doing, the image of the Soviet state is less tarnished than in other republics (Ukraine, for example). The story of Kurapaty is an important event because it demonstrates how firmly the Belarusian republic had been placed under state control. Though similar massacres have been uncovered in most other Soviet republics, only in Belarus have the authorities managed to rewrite history once again in their favor. The fact that this revisionism was achieved at the end of and even beyond the Soviet period may define the limits of the democratic movement.

On 3 June 1988, the discovery of a burial site at Kurapaty, some two miles north of Zelyoniy Luh, was reported in the organ of the Belarusian Writers' Union, *Litaratura i mastatstva*, by Zyanon Paz'nyak and Yaugen Shmyhaleu.[45] The article opened with an emotional account of some of the more horrific events of the Stalin purges, with Paz'nyak expressing regret that he had not recorded events related to him by those returning from the Gulag camps. However, many elderly witnesses had provided him with information about the events at Kurapaty. They related that two miles to the north of Zelyoniy Luh village, there was an execution site, at which shootings occurred on a daily basis in the period 1937 to 1941. They also declared that an old forest was located there, a considerable portion of which (10–15 hectares) was surrounded by a wall more than three meters in height. Paz'nyak had learned of these facts in the 1970s, but at that time 'there was no possibility of revealing the truth to the world.'

The government was obliged to respond to these revelations, particularly after a mass meeting at the burial site later in the summer of 1988. A Government Commission was formed, though it was notable for its absence of what might be termed impartial witnesses. In addition to party, government and KGB officials, it also included a number of former partisans whose attachment to the Soviet regime had remained quite rigid. Paz'nyak himself denounced the subsequent report as a 'whitewash.' The report was badly flawed and was soon assailed by Communist sources, such as the journal *My i vremya*, which expended several columns elucidating the loopholes in the rapidly assembled government account. The outcome was that within a short period, the Kurapaty massacres were attributed to the Germans. Documents on the subject were not made available outside the office of the public prosecutor. Further, Paz'nyak himself, an emotional archeologist, may have inflated the totals of bodies discovered at the site.[46] Remarkably the uncovering of a Stalinist massacre – hardly uncommon over the past seven years – had been followed by a government-inspired cover-up. Kurapaty, however, was hardly the only such site.

In November 1994, for example, a new 100-bed oncological dispensary was under construction in Vileyka (Minsk Oblast), an area incorporated into the USSR in September 1939. On 22 December when a ditch was dug for the foundations of the new building, a mass burial site was found. As the builders dug further, more bodies were uncovered. In March 1995, when the information about the site was revealed publicly, it was announced that the Vileyka site had revealed the remains of a mass execution site of Soviet citizens by the Germans in 1941–4. All other possible variations of the event were denied. An alternative explanation has, however, been furnished by the Belarusian historian, Igor Kuznetsov.[47] Kuznetsov recalls that in September 1939, the advance units of the Red Army arrived in Vileyka, followed by NKVD squads that at once began to identify potential enemies among the newly-annexed population. These included former members of the Communist Party of Western Belarus. The NKVD rounded up some 854 prisoners out of a total of 910 in Vileyka prison alone in May 1940 in addition to many others from different localities. Between September 1939 and June 1941, at least 1000 of these prisoners were executed. Eyewitnesses have recalled the shootings of these 'enemies of the people.' They were led to a special dugout, strangled with a leather strap and then hurled into a hollow under the prison walls. Before the invasion of the Germans, hundreds of Vileyka residents were reported as 'missing.' Many had been executed by the NKVD as late as 22 June 1941, only a matter of days before the arrival of the invading army. In short, therefore, the Kurapaty massacre has been supplemented by the Vileyka executions despite the obvious reluctance of the authorities to make the logical deduction of Soviet responsibility.

The Incorporation of Western Belarus

The 1930s reversed any political gains the Belarusians might have achieved in the tolerant 1920s and ensured that the BSSR would be ruled by compliant leaders who would rigorously follow Moscow's line. Not until the advent of Masherau as the leader of the Communist Party of Belarus in 1965 did the authorities demonstrate any semblance of cultural awareness. One could argue that Ukraine to the south was dealt with in a similarly ruthless manner. However, the political and cultural organizations in the Ukrainian areas formerly under the Austrian empire and transferred to Polish rule in the interwar years were far more advanced than their western Belarusian counterparts and contained a much larger proportion of the indigenous group. By the mid-1930s, western Belarusians had no representation in the Polish parliament; and most Belarusian cultural organizations were disbanded in 1936–7.[48] These actions, combined with the authoritarian nature of Polish rule and its deliberate polonization of the Belarusians – particularly the Catholics – deprived them of an alternative source of cultural retention and political awareness. By the late 1930s, however, Poland's hold over this region was becoming increasingly weak. On the other hand, there was in this same period little scope for pro-Communist activity because the Communist Party of Western Belarus, as a composite part of the Communist Party of Poland, had been dissolved by the Comintern on Stalin's orders in July 1938.[49]

As a result of the Molotov–Ribbentrop Pact of August 1939, the western regions of Belarus were assigned to the Soviet sphere of interest in preparation for the German-Soviet dismemberment of Poland. In Soviet parlance, the invasion was officially an act of 'liberation,' though the phrasing of such terms had to be directed against the former Polish state rather than the very real menace of Hitler's Germany. Thus the Red Army had intervened to protect its 'blood brothers,' the western Belarusians, who had for years languished beneath the 'yoke of landlord Poland.' Led by General M.P. Kovalev, the Red Army soon took over Belarusian territories of eastern Poland. An 'election' to a People's Assembly was held (96.7 percent of the electorate took part) and a single slate of candidates representing 'workers, peasants and toiling intelligentsia' was overwhelmingly elected. The Assembly convened in Bialystok on 28 October 1939, and ended its activities two days later with an appeal to the USSR Supreme Soviet to be incorporated within the Soviet Union. On 12 November Western Belarus was formally incorporated as part of the Belarusian SSR.[50]

The incorporation of western Belarus (today's Brest and Hrodna oblasts and the western part of Minsk Oblast) in 1939 and the invasion of Belarus by Nazi Germany heightened national problems. Western Belarus was regarded by the Soviet authorities as an area of potentially dangerous 'bourgeois nationalism,' and the period 1939–41 was notable for its harshness and mass deportations, and for executions of western Belarusians at sites such as Kurapaty. Kabysh estimates that by the end of February 1940, 140 000 western Belarusians had been arrested and deported from their homeland, and notes that a second wave of arrests followed in May

1940, encompassing a further 70 000 people; a third occurred shortly before the German invasion in June 1941.[51] To put these figures into perspective, the population of western Belarus was about 4.8 million, though this figure included the Poles, many of whom fled from the invading Red Army or were arrested at an early stage of the occupation.

The question has been debated whether, as in Ukraine, national consciousness in Belarus was enhanced by the annexation of ethnic territories that had lain outside the Soviet domain. The percentage of ethnic Belarusians incorporated was considerably lower than the number of ethnic Ukrainians in Polish territories to the south. For example, in a penetrating article on the subject of the incorporation of Western Belarus, Mikolaj Iwanow points out that in terms of territory, the annexations were much more significant for Belarus than for Ukraine, resulting in a 45 percent increase in total area and a population rise of 46 percent (5.6 to 10.4 million). Iwanow cites also the 1931 census and revised figures of the Polish historian Jerzy Tomaszewski which would indicate that the ethnic Belarusian population in western Belarus was between 21 percent and 40 percent of the total population. A Soviet source maintains that some three million Belarusians were 'liberated' in the western territories in September 1989, but the figure appears inflated.[52] The incorporation of the region and the 'reunification' of ethnic Belarusian territories into a single republic was a blatant propaganda campaign to attribute a new unity to the socialist system and Stalin in particular.[53] Repressions against western Belarusian Communists had begun in the 1930s, culminating in the dissolution of the Communist Party in 1938 as noted above. The repression of the nationally conscious elements in the newly annexed territories began alongside the ruthless measures directed against the Poles – indeed, according to Iwanow, 'the soul of the nation' was destroyed, beginning in Vilna.[54] Even a Communist source notes that progress in the new territories was greatly hindered by 'gross violations of legality and social justice caused by the cult of personality.' Tens of thousands of citizens were repressed, particularly former members of the Communist Party of Western Belarus.[55]

The annexation of western Belarus raised the possibility that the city of Vilna – which can accurately be described as the cultural repository of the Belarusian nation, though Belarusians resident there constituted only a small minority of the population – might once again be included within the borders of a Belarusian state, albeit under Soviet auspices. Indeed, the Soviet authorities prepared for such an inclusion with the establishment of a Belarusian-language newspaper in this city called *Vilenskaya pravda*. However, on 10 October 1939, the city, along with the Vilna region as a whole, was transferred on the orders of Stalin (in agreement with the

Germans) to Lithuania.[56] Molotov noted that the majority population of the city was not Lithuanian but maintained that the handing over of the city was more of a moral question (!) and satisfied Lithuanian aspirations.[57] The impact of this decision was to solidify the importance of Minsk in the BSSR and to allow Minsk to replace Vilna as the cultural center of the republic, a city that was already firmly under Russian cultural hegemony. Had Vilna become part of Belarus, one might have witnessed a simultaneous development of two rival cities that could have significantly altered the political landscape of the country. By comparison, one can observe the impact of the inclusion of the city of L'viv into the Ukrainian SSR, a city that remained an outpost of national consciousness and a center for dissidence in the 1960s. Belarus was to be given no such options and the decision was ultimately to affect the formation of the post-Soviet state.

World War II

As noted earlier, the impact of World War II is a complex one that requires a separate treatment. The intention here is to examine its influence on the development of the republic. Belarus was occupied in remarkably rapid time by the German army. Within six days of the German invasion, the capital city of Minsk had been occupied.[58] The official Soviet portrayal of valiant resistance at all stages of the campaign is belied by the total collapse of the defensive system. That there was some support for the invader among the population is evident. In contrast to some occupied areas, this support existed at a relatively late stage in the war. On 27 June 1943, for example, by which time the nature of Nazi rule was evident to the populace, the invaders belatedly attempted to establish a puppet regime with the organization of a 'Council of Men of Confidence' under the leadership of a former Slutsk landowner, Radaslau Astrouski. By the end of the year, the Germans had also conceded the formation of a Belarusian Central Council, of which Astrouski was the president. Astrouski demanded that a Belarusian national army be formed and used exclusively against the Bolsheviks.[59] The national flag and symbols were restored by this short-lived 'government,' a factor that was to have repercussions for the independent state 52 years later when the new national symbols were associated with collaborationist sentiment.[60]

The extent of the anti-Soviet movement in Belarus is one of the more difficult factors to assess. Clearly the Stalin regime had alienated a segment of the population. Conversely the success of the partisan movement in this area was exaggerated by Soviet historians, in terms of both numbers and the period of influence.[61] Indeed there is little evidence to indicate that the partisan movement was significant prior to 1943. According to one source, in that year, some 96 000 people joined the partisan ranks, so that by November, there were a total of 122 600 partisans.[62] Based on these figures from the Khrushchev era therefore, over 78 percent of the partisans began their activities in 1943 and, judging by the date provided, late in that year, at a time when the ultimate result of the German–Soviet war was no longer in doubt.

The above statement is not intended to question the resilience and achievements of the partisans, which are beyond question. The intention is to emphasize the hyperbole that continues to surround the movement. Belarus suffered greatly from the war. A quotation from Heinrich Himmler, SS Reichsfuehrer, that is dated 1940, declares the intention to either exterminate or dispatch 75 percent of the Belarusian population to Siberia, and subject the remainder to Germanization for use as a labor force.[63] During the three-year occupation, Belarus was the site of several concentration camps, including that at Trostyanets, near Minsk, the third largest in the German-occupied territories of eastern Europe.[64]

According to Soviet sources, the occupation forces were responsible for the deaths of 2.2 million residents of Belarus, while a further 380 000 were sent as *Ostarbeiter* to Germany (mainly women and young adults). Industrial losses included the destruction of 10 388 factories and enterprises, 96 percent of the republic's energy capacity, and almost 80 percent of equipment of the building materials industry.[65] Officially also, the Germans burned over 500 000 collective farm buildings, destroyed virtually the entire stock of animals, some 7000 schools, almost all theaters, clubs, museums and libraries.[66] Such an orgy of destruction is hard to imagine and it is conceivable that the large-scale battles on the territory of Belarus would have accounted for some of the losses cited above. Yet the savagery of the German occupation regime was responsible for what might be termed 'new patriotism' among the Belarusian population and the resolute nature of both partisan activity and the costly advance of the Soviet army in general. Henceforth – and it is posited that this occurred at a relatively late stage of the war – the causes of the Belarusians and the USSR were united. Belarusians associated themselves with the patriotism invoked by Stalin two years earlier at the November 1941 commemoration of the October Revolution.

The advance against the Germans occurred on a vast 1100 kilometer front and involved four armies, namely those of the First Baltic Front commanded by General I.Kh. Bagramyan, and the First, Second and Third Belarusian Fronts commanded by Generals K.K. Rokossovskiy, G.F.

Zakharov, and I.D. Chernyakhovskiy. In the spring of 1944, there was a widespread partisan attack on the Germans in the 'Belarusian Military District.'[67] However, the most significant period from the Belarusian perspective was 26 June to 28 July 1944, which saw the recovery of all Belarusian territory, highlighted by the recapture of Minsk on 3 July and culminating in the liberation of Brest by the army of the First Belarusian Front on 28 July. The rapidity of the Soviet advance paralleled that of the German army in the summer of 1941. Though losses were high, the Red Army was determined to cross the former Soviet border and enter Poland at the earliest opportunity.

For Belarus, the wartime era was a significant one. Though doubtless with other motives in mind, Stalin was justified in emphasizing the sacrifice made by the republic in the war years at the Yalta conference. The outcome was Belarus's advancement to membership of the United Nations on 30 April 1945 and participation in the San Francisco conference in that same period.[68] While this acknowledgment might be dismissed as ceremonial, given that neither Belarus nor Ukraine ever showed any inclination to differ from Soviet directives in the UN, for the Belarusians themselves, the new position was further testimony to the sacrifices made during the war. The Great Patriotic War became the touchstone of the BSSR's existence. Its wartime heroes became its postwar leaders.[69] The war dominated the psyche of the population and the subsequent Communist leadership. By the same token, the eventual emergent Belarusian nationalism of the 1980s and 1990s could never quite eradicate the stain of collaboration, whether or not such a link was ever justified. The Soviet authorities would always henceforth attempt to associate national culture and development with 'bourgeois nationalism' and the crudely phrased 'Belarusian-German nationalists.' Because of the significance of the war in the republic, the association remained.

In World War II the losses in the republic were proportionally higher than for any other single state in the world, with approximately one in every four citizens falling casualty to the occupation regime, or as a result of the battles between the German and Soviet armies. The wartime destruction of the Belarusian villages and the goals of the all-Union Fourth Five-Year Plan (1946–50) to concentrate primarily on industrial expansion rather than rural recovery signalled the route taken in the Belarusian SSR to develop and invest in industry to the detriment of the countryside. By the end of the plan period, industrial output in the republic already exceeded the prewar level by 16 percent, as a result of the development of machine building, energy production, the chemicals industry and other sectors of the economy.[70] Thus the events of the war catalyzed the demographic processes of the postwar era. The village lacked amenities, roads, educational and cultural establishments and was regarded by the authorities as little more than a supplier of goods to the town.[71] For advancement in life, the enterprising citizen had to move from village to town. This is hardly a unique process in the postwar USSR, but it took on more extreme forms in Belarus. As one Soviet source acknowledged with rare frankness:

> However, the situation in agriculture in the republic [in 1946–50] was still far from satisfactory. Many party and soviet organizations weakly led agriculture, and soviet organizations failed to procure a high tempo for the reestablishment and development of collective farm production. The majority of farms insufficiently used the achievements of science, and the experience of progressive workers was introduced weakly. On many farms technology was used inadequately, and there were gross defects in the organization of labor.[72]

Belarusian agriculture, in truth, never recovered from the effects of the war, and state policy which neglected its welfare to an extraordinary degree. For the authorities, postwar recovery lay in the development of the city and in city industries. Belarus's future had been clearly mapped out as an industrial repository of the Soviet borderlands.

POSTWAR URBANIZATION

This growth of large urban centers became more accentuated after World War II.[73] Indeed by the 1960s, even the Soviet authorities felt obliged to address the question of whether cities in the republic were being developed too rapidly. Thus in a 1990 article, Vasiliy Kharevskiy pointed out that because of the 'extraordinary increase' in the rise of large cities, and the concomitant slow development of small cities, the republic was facing a mass extinction of villages and hamlets. Kharevskiy expressed particular concern about the growth of large factories and other enterprises that tended to concentrate in already large conurbations rather than small ones. The city of Mahileu in particular epitomized the problems of urbanization with its chronic shortage of residences, poor road networks and acute environmental problems.[74] In the period 1959–86 therefore, while the total population of the Belarusian SSR rose by 24 percent; that of urban centers rose by 250 percent, and by 1990, 66 percent of the population was located in cities as compared to 31 percent in 1959.[75] Of the total urban population in the late 1980s, 60 percent lived in cities of more than 100 000 population.[76]

The industrialization and urbanization of the Belarusian SSR resulted in several distinctive features in the composition of the republic. First, as noted, Minsk inherited from Vilna the position of the leading city, though they had been rivals in the past. After the 1921 Treaty of Riga, Vilna and other territories of western Belarus were included in the reestablished Polish state, which in practice if not in theory did not recognize any claims to autonomous status in its Belarusian territories. The loss of Vilna in fact also left Minsk as the only major urban center in the new Soviet republic. The dominance of Minsk even compared to other republics was especially notable. In few other republics was there a capital city so centrally placed and which outpaced its rivals so markedly in growth. By 1939, the population of Minsk was 237 000 compared to 167 000 in Vitsebsk (the nearest rival), and 139 000 in Homel'.[77] By the end of the Soviet period, however, Minsk was three times larger than Homel', then the second largest city. Table 1.2 illustrates the remarkable growth of the capital city compared to its neighbors in the BSSR.

Table 1.2 indicates that the growth of Minsk kept pace with the rise in the urban population generally in the postwar years, and that the city was the residence of almost one-quarter of Belarus's urban residents. If one compares the percentage of urban residents living in Minsk and Homel' between 1959 and 1989, one can see that this percentage remained fairly stable throughout the Soviet period (Table 1.3).

By the 1960s roughly one-sixth of total residents (urban and rural) of Belarus lived in the capital city (even today that percentage has only declined slightly). Minsk was also the center of educational life, publishing and printing, and culture, in addition to being the administrative

Table 1.2   *Growth of major Belarusian cities in the postwar years (population in thousands)*

City	1959	1970	1979	1989
Minsk	509	917	1276	1613
Homel'	175	282	394	512
Mahileu	122	202	290	359
Vitsebsk	148	231	301	357
Hrodna	73	132	195	271
Brest	74	122	177	258

Source: State Committee of the Belorussian SSR for Statistics, 1990, pp. 15–16.

Table 1.3   *Proportion of urban residents living in the cities of Minsk and Homel'*
in the postwar period

	Minsk	Homel'
1959	20.5	7.0
1970	23.5	7.2
1979	24.2	7.4
1989	24.1	7.7

Source: Calculated from ibid.

capital and headquarters of the Communist Party leadership. The dominance of Minsk may have enhanced the national self-awareness of
Belarusians as a national group, but it also stifled national expression, pre-
venting it from becoming a dominant factor in political and cultural life.
As in the Stalin period the city was also largely Russophone and a policy
of expansion of Russian cultural and language publications was soon to be
inaugurated. This situation perhaps embodies the essence of the Belarusian
identity crisis that persists today: how to equate pride in national achievements – such as industrial progress or culture – with the Russification of
the dominant urban center of the republic. One can posit that such an
unnatural development in a republic was not necessarily planned, at least
not to such an extreme degree, though it may have been the natural
outcome of Soviet economic policies. Before we examine this subject in
more depth, let us also take into consideration other features of state
development of the early Soviet period.

2 Language and Culture: National Nihilism?

THE PLIGHT OF THE VILLAGE

The Stalinist system favored the growth of large urban centers for the development of heavy industry: they were also more convenient for dissemination of propaganda and the growth of party and Soviet organs. Belarusian national consciousness, on the other hand, suffered as a result of urban development. All the large cities in Belarus (Hrodna being a possible exception) after the 1930s were repositories of the Russian language and the Belarusians were more easily assimilated in an urban environment. Although a small indigenous Belarusian elite may have had the opportunity to emerge and play a role in the political process, it was also dependent on the vigorous growth and preservation of a rural culture. Belarus's form of national development as a major industrial republic in which national consciousness has lagged behind that in its neighbor states owes much not only to its demographic development, but also concomitantly to Soviet language policy.

The Belarusian scholar G.I. Kasperovich illustrated this process in a notably frank article in 1985. He observed that the city life style compared to village existence plays a major role in the formation of a unique Communist way of life. Cities also play a part in cultural exchange between the peoples of the USSR and the dissemination of internationalist views, all of which was considered laudatory. However, he added, there are some dark sides to this process. For example, in cities the process of 'natural language assimilation' was increased, and there was wider application of the languages of international usage 'which for our country [signifies] the Russian language.' By contrast, the village was the carrier of folklore and traditional culture; the songs and art of Belarus; in addition to the native language.[1] The situation in Belarus, in his view, was brought about less by the Soviet authorities than by past waves of migration that brought Russians and other groups into Belarusian territories (in the case of the Russians in the late-eighteenth and nineteenth centuries). But in the Soviet period, the virtual abandonment of the villages was, implicitly, tantamount to the destruction of the local Belarusian folk culture.

This feature of the Soviet period can be exemplified in several ways. One can perceive a distinct change in the size of families in the period 1970 and 1989, and the deterioration of family life in the villages. In 1970, rural families were slightly larger than their urban counterparts, averaging 3.7 members to 3.5. By 1979, the size of urban and rural families was about equal at 3.3 members. In 1989, however, the average size of the village family had dwindled to 3.0, while the urban family still averaged 3.3 people.[2] The statistics provide evidence of the depopulation of the Belarusian village. The percentage of the rural population in the overall population of the republic has fallen from 43 in 1980 to 33 today. Out of 10.2 million people living in Belarus in 1990, only 3.37 million lived in rural regions.[3] Perhaps the young people would find it convenient to move to towns for better jobs and education? If this were the only demographic problem facing Belarus, the dilemmas would surely be surmountable. There is evidence, however, that what Mal'dis terms national nihilism can be applied to young people generally. The majority of them have little faith in the future of their republic. In one sociological survey, 71 percent of the young people interviewed declared that they would be prepared to emigrate to obtain a better career. What is termed the 'emigrant mood' evidently encompasses the socially active part of the population: students, aspiring diplomats, highly qualified specialists, sports and entertainment figures.[4] The lack of faith in one's nation appears to be endemic in the Republic of Belarus. Thus, not only does one find the villages depopulated, but the city life has not provided satisfaction to the young people of the newly independent state.

The depopulation of villages was paralleled by a general decline in the growth rate of the total population in the republic in the period 1965–75 in particular. The cause was said to be a rapid fall in the birth rate coupled with an almost stable mortality rate. Several reasons have been postulated for this situation, such as economic phenomena: a shortage of labor that often entailed demographic changes, such as the separation of husbands and wives. In 1970, 40 percent less children were being born than in the 1950s. In turn, the ageing of the population was a major cause of the decline in the birth rate. The percentage of the population over the age of 50 has risen steadily throughout the twentieth century on Belarusian territories, from 12.6 percent in 1897, to 13.9 percent in 1939, 20 percent in 1950, and 21.2 percent in 1970.[5] According to the census of 1989, the over-50 group has increased further to 28.7 percent.[6] The natural increase of the population of Belarus has dropped from 17.8 percent annually in 1960 to 6.1 percent in 1980, and remained steady at 1.7 percent in 1990 and 1991.[7] Aside from the disappearance of villages, therefore, the republic is approaching a zero and even a negative growth rate.

As an example of the pervasiveness of 'national nihilism,' it is instructive to turn to the language question in Belarus in the Soviet and post-Soviet era. There, perhaps more than in any other facet of life, the Belarusians might be termed unique among the former Soviet republics in that the majority of the urban native population does not speak its mother tongue. Epitomizing this dilemma is the fact that the country's first president may be the only figure in modern history to hold so eminent a position and yet be unable to communicate effectively in the native tongue.

THE LANGUAGE QUESTION

It has been pointed out by several scholars that although Belarus lacked national statehood, the Belarusian language received a prestigious position as the language of communication in the Grand Duchy of Lithuania. It was used for daily business, official and church statutes, and in chronicles.[8] After the union of Lithuania and Poland in 1569 at Lublin, however, Belarusians were subject to Polish influence and the Roman Catholic religion. After the partitions of Poland in the late eighteenth century, most of the Belarusian territories once again were incorporated into the Russian Empire. Demographically, ethnic Belarusians made up most of the peasant stratum, engaged in subsistence agriculture, though a few had penetrated the higher professions (10 percent of the lawyers and 21 percent of teachers in Belarusian territory were ethnic Belarusians). According to one source, over 75 percent of ethnic Belarusians in the age group 10–49 were illiterate.[9] Shabailov maintains that for tsarist Russia, the Belarusian territories never constituted more than a region, with no recognition of a Belarusian people. In pre-revolutionary Belarus, he states that up to 80 percent of the population was illiterate.[10] Another source states that after 1795, the Belarusian language was deprived of all rights on Belarusian territories. From 1864 until 1905, no Belarusian-language publications were permitted by the Russian monarchy. Educational establishments were closed, beginning with the University of Polatsk in 1830.[11] Thus the Belarusian language would appear to have reached a low point in its usage and recognition on the eve of the Soviet period: it was confined to the countryside, and to a largely illiterate population.

As noted briefly above, illiteracy in the BSSR was eliminated during the Soviet period, a process that may be considered a significant accomplishment. According to one source, 68 percent of the population living in what today comprises the Belarusian state in the pre-Soviet period was illiterate; yet by 1920 this figure had been reduced to 52.6 percent. The campaign to eliminate illiteracy in 1924–5 reportedly included the services of over 65 000 people, based predominantly in the rural localities.[12] By 1926, during the period of renewed cultural development, only 40.3 percent of the population was termed illiterate.[13] Ivan Lubachko notes, for example, the attempt to recultivate Belarusian as a literary language after three centuries of non-usage. He observes further, renewed interest in distinguished Belarusian cultural figures of the past, such as Frantsishka Skaryna and Symeon Polotski. The latter figure was particularly acceptable to the authorities as an advocate of a Russian-Belarusian union as opposed to Polish control over the territories of Belarus.[14]

The Soviet authorities in Belarus conducted school reforms after 1924, and raised the annual school budget from 5.4 million rubles in 1923–4 to 14.8 million rubles in the 1925–6 year. The key problem at that time was the lack of adequately qualified teachers as 26 percent did not have a middle education.[15] However, the period was a significant one for the development of Belarusian culture. Talented exiles were encouraged to return home; the Belarusian language was being used in elementary, middle and high schools; the language was also finding its way into party and Soviet publications; and was becoming the principal language in the republic for the first time in the Soviet period. The famous Belarusian poets Yanka Kupala and Yakub Kolas, now in their declining years, took part in this process. However, the key question is whether this cultural renaissance could be maintained once the more tolerant years of the New Economic Policy ended. By the late 1920s, the period of tolerance was over. The early Stalin years saw an assault on the peasantry that brought with it a reversal of the cultural policies hitherto developed. Henceforth, proponents of the Belarusian language would be deemed 'bourgeois nationalists' and the movement toward cultural progress 'national deviationism.' Belarusian writers and scholars were accused of separatist sentiments and of inventing a 'counter-revolutionary orthography.'[16]

The decline of the Belarusian language in the Soviet period can by no means be depicted as a simple downward curve. It experienced early growth and development, harsh repression, fluctuations, periods of revival (particularly in the period of Masherau's leadership of the Communist Party of Belarus), and official concern. In the early postwar years, Belarusian literature was encouraged to focus on themes related to the war, and the period 1946–58 also saw the publication of a significant quantity of children's literature on this theme in the native language.[17]

Simultaneously, however, the school system became more heavily politicized, and the year 1946 was notable for a massive purge in Belarus that encompassed 90 percent of regional party leaders, 96 percent of soviet officials at the raion and city level, and 82 percent of collective farm chairmen. During this process, the Belarusian leadership structure was Russified even to the extent of the appointment of a native Russian, Nikolay Gusarov, as First Party Secretary.[18] One can make the case that in terms of mere numbers the development of schools, higher educational institutions, theaters and museums significant progress was made in the cultural sphere in the period 1945 to the late 1960s.[19] For Belarusians, however, the development of their language and culture was tempered by harsh political administration and the need to emphasize the fraternal role of the Russians in the reconstruction of the republic after the war years.

In western Belarus, which had suffered German occupation for the longest period, a key concern was the area's subjection to 'bourgeois influences' both prior to and during the war. The Soviet authorities also considered this area to be the prime repository of a 'national culture' and to merit serious concern, even though, as we have argued above, the Belarusian element within the population was never a dominant one in the 1920s and 1930s. Consequently, western Belarus was subjected to harsh collectivization of its agriculture, and 'hostile' elements – defined as 'kulaks,' 'Polish and Belarusian bourgeois nationalists' and the 'reactionary Catholic clergy' – were systematically destroyed. The Soviet version of these events illustrates indirectly the scale of the opposition to the Stalinist regime. After the territory was liberated and the Germans had retreated, bands of activists remained behind which assassinated party and soviet workers, and terrorized the population. The region therefore was collectivized with the use of the political sections of the Machine-Tractor Stations (MTS), the key instrument in the brutal collectivization campaign of the early 1930s, and 6000 new collective farms were created in Belarus by the end of 1950.[20] Under these circumstances, the distinctiveness of western Belarus within the Soviet republic was eliminated.

By the late Soviet period the question of the Russification of culture in the republic had become so critical that it led to some inquiries and self-searching among the intelligentsia. In particular, the decline of the Belarusian language had reached an acute stage because many Belarusians were conversant only in Russian and used the Russian language as the basis for their daily activities. The dilemma was described eloquently by Sobolenko:

The change of language certainly does not lead automatically to a change of national consciousness, but nevertheless it often precedes it. And if language assimilation is not always complemented by ethnic assimilation, then ethnicity without language usually does not exist.[21]

In short, therefore, the survival of a national republic is, if not dependent upon, at the least much enhanced by the flourishing of the native language and culture, particularly one with such a long history as the Belarusian language. If the language is no longer used by the native intelligentsia and this group has effectively become assimilated to the use of the Russian language, what does this signify for the preservation of a national state, particularly one that is already heavily dependent economically upon its Russian neighbor, and most of which has a lengthy history of Russian rule with its varying degrees of repressiveness in the tsarist period?

As Belarusian cities, and especially Minsk, grew dramatically in size in the Soviet period, the language question became more acute. A language crisis evident in the tsarist period was exacerbated in the Soviet era, when the vast majority of publications in the BSSR were in Russian rather than Belarusian, and Russian became the language of the urban workforce. Sobolenko also pointed out that the younger generation has been far more prone to become assimilated to the use of the Russian language than the older generation. He noted that among Belarusians over 60, only 3.6 percent declared Russian to be their native language in the census of 1970, whereas among those aged 30–39, the figure was 9.4 percent, and in the age group of 16–19 years, almost 14 percent.[22] The language crisis in the Soviet period began to attract the attention of concerned Belarusian scholars after the census of 1959, the first Soviet census in 20 years. Thereafter, it is possible to delineate clearly the decline in native language usage.

In the period between the censi of 1959 and 1970 there was a significant decline of Belarusians in the republic with native language fluency (78.99 percent – 76.64 percent) and a notable increase of those with Russian as their mother tongue (16.26 percent – 21.19 percent). The number of Belarusians speaking their native language also rose much slower than the natural growth of the indigenous population. Most of the gains for the Russian language in this period were in Minsk.[23] In 1959, the number of declared Belarusian and Russian speakers in this city was about equal: 247 922 Belarusian speakers to 246 276 Russian speakers. By 1970, Russian speakers composed 54.5 percent of the population, or 104 000 more than Belarusian speakers. Most of the rise in Russian speakers was not a result of an influx of ethnic Russians into the capital city, but rather because of the adoption of the Russian language by Belarusians themselves.[24] An

American scholar, Roman Solchanyk, has noted that language retention among Belarusians in the 1970 census was the lowest among the 15 national republics of the USSR. In addition, almost 68 percent of those who declared Belarusian to be their native language in 1970 were also fluent in Russian.[25]

Further perspectives can be gleaned from a 1985 book edited by A.E. Mikhnevich, which points out that according to the 1979 census, 5.1 million people, or 53.6 percent of all residents of the BSSR, spoke Russian fluently, while 2.7 million considered it their native language. Of that 2.7 million, about 45 percent were native Belarusians.[26] Thus 82 percent of republican residents had complete fluency in Russian and clearly the percentage in urban regions was much higher, particularly in the city of Minsk. These figures generally may represent a significant underestimate. In a speech at the May 1979 Tashkent conference entitled 'The Russian Language – the Language of Friendship and Cooperation of the Peoples of the USSR,' M.G. Minkevich, the Minister of Education of the BSSR, declared that over the years of Soviet rule, Russian had become the second native language of every Belarusian, and that there was scarcely one person in the republic who did not have some knowledge of Russian. About 61 percent of all school pupils were at that time being taught in Russian.[27] The late Brezhnev period may thus be regarded as something of a heyday for the growth and development of the Russian language in Belarus.

Zyanon Paz'nyak, head of the Belarusian Popular Front, historian and archeologist, has been among the most eloquent proponents of the use of the native language. He also has lamented the events of the 1970s. By the middle of that decade, he has noted, some 40 years after the beginning of the 'linguistic experiment', none of the 95 Belarusian cities and only a handful of the 117 raion centers possessed a Belarusian school; had a Belarusian kindergarten. Even in villages and hamlets, Belarusian schools had been closed down. Though Minsk had a population of 1.5 million, it contained only Russian-language schools. In the entire postwar period, he continued, not one teacher had been trained for a Belarusian school. Teaching at all higher educational institutions and technical colleges was exclusively in Russian. All business affairs from collective farms to the highest echelons of the government were conducted in Russian.[28]

Once again, however, it is important to focus on the role of the capital city in this process. In his pioneering article, 'The Belorussians: National Identification and Assimilation,' Steven L. Guthier notes that the 'decline of the native language parallels the period of intensive population growth in the capital.' He points out that in the period 1959 to 1973, whereas the urban population of Belarus as a whole grew by 69 percent, that of Minsk increased by 104 percent.[29] By the late 1960s, the capital city was the publication center for a formidable output of Russian-language works. In Belarus as a whole, by 1967, of the total of 1.8 million books published, only 337 000 (18.8 percent) were in the Belarusian language, and in terms of copies in circulation, the percentage was 37.6 percent. The situation with publication of newspapers was deceptive. Of a total of 176 published in the republic in 1967, 133 were in Belarusian. However, in terms of copies in circulation, only 38.7 percent were Belarusian-language newspapers and 61.3 percent Russian language.[30]

According to Paz'nyak, what might have been even worse than loss of language was the development of a so-called 'half-language,' a mixture or patois of Belarusian and Russian which was spoken, he estimated, by 50–60 percent of the Belarusian population. It was referred to as '*trasnyaka*' and was particularly endemic among the middle and lower social strata of the population. Only a portion of those who spoke this language were fluent in Russian and they did not speak Belarusian. The use of this patois was reportedly having a very adverse effect on the development of Belarusian culture. One of the main problems, in Paz'nyak's view, was that people were confused; it was impossible for them to develop and to exist simultaneously with two cultures. A paradox had been created. Estonians and Latvians from various walks of life, who were essentially unilingual, still spoke Russian better than those in Belarus who spoke this patois, but lacked a fundamental knowledge of the structure of the Russian language.[31]

The situation did not improve significantly in the Gorbachev period. One can turn to the example of Hrodna Oblast, an area of the republic in which Russification was at a relatively low level in the late 1980s. In the city of Hrodna at this time, there were reportedly 143 000 citizens of Belarusian ethnicity, 59 000 Russians and 58 000 Poles, in addition to other ethnic groups. The city party committee had appointed a commission to study the question of fulfillment of the oblast program entitled 'Native Language,' which was geared toward the implementation of Belarusian as the state language. After a month the commission provided the following data: of the 143 000 Belarusians in Hrodna, 60 000 considered Russian to be their native language. Only 15 percent of Belarusian parents felt that their children should study at an exclusively Belarusian school; 45 percent were categorically opposed; and the remainder favored the devotion of an equal amount of study time to the two languages. Only 8.3 percent of elementary schools were currently providing Belarusian language study.[32] One can confidently assert that if the above represented the situation in

Hrodna, then even more difficult circumstances for Belarusian language instruction and adoption existed in Vitsebsk, Salihorsk and other cities.

By 1989, the Belarusian scholar A. Mal'dis was describing what he termed a 'national nihilism' which was evident in all union republics, but particularly endemic in Belarus. The most eloquent indicator of this phenomenon was, in his view, the mass repudiation by Belarusians of their own native language so that the city population, including virtually all the middle stratum, employees and technical intelligentsia, did not speak Belarusian. The situation in 1989 had become so acute, Mal'dis pointed out, that in certain oblasts, such as Vitsebsk, and in the vast majority of raion cities, there was not a single Belarusian school. In those areas that did possess schools, the rural raions were woefully lacking in teachers of non-philological subjects, and few were capable of communicating in their native language.[33]

The chief problem, however, in Mal'dis's opinion, was the reaction of the authorities. Either they had ignored the problem or else they had attacked the whole concept of nationalism and national development. In many documents and articles – he singled out as especially harmful and short-sighted items published in *Vecherniy Minsk* and *Za peredovuyu nauku* – it had been stated that the only real danger to the republic was nationalism. National nihilism, expressed in such articles, had to be perceived as a malaise that should neither be exaggerated or underestimated. The people's poet of Belarus, P. Panchenka, had declared the national losses to be irreplaceable, but Mal'dis declared himself to be more optimistic. The situation might indeed have been irreversible, he added, had there been no perestroika, or if perestroika had occurred twenty years later.[34]

According to Mal'dis, the national misfortunes of Belarus dated back to the second half of the sixteenth century, prior to which national development could be termed 'more or less normal.' In pursuit of privileges, he notes, feudal leaders began to adopt the Polish language and Polish culture. The old Belarusian language, which had been the state language in the Grand Duchy of Lithuania, was ousted from the official sphere by the end of the seventeenth century, and restricted to household use. When Belarusian lands were transferred to tsarist Russia, the upper stratum of society, 'with a carefree attitude,' switched to the Russian language. Consequently, at the beginning of the twentieth century, the Belarusians did not constitute a nation. Though some beginnings had been made in the Lenin period of the USSR, the Stalinist repressions, 'when everything national was equated with the nationalistic,' and the stagnation period,[35] when 'national nihilism' was in vogue, developed into a compulsory regimen. The disease was a 'chronic sickness,' the cure for which did not promise to be easy. In his view, the answer lay in the system of higher education: the preparation of national cadres for schools and cultural institutions. Teaching in Belarusian should ideally begin in the second or even the first grade. However, who was to perform such a task? Few teachers were capable of teaching in the native language. In Minsk, and especially in other oblast and raion centers, there was either a dearth of Belarusian-language teachers, or else those who did teach the language were not well versed in it – and this problem had reportedly traumatized the children. Mal'dis stressed that the solution must begin with the family, but the family had forgotten the language and was unlikely to rekindle its usage until it had become necessary, prestigious, used for communication in the workplace and school, and used in speeches in the Supreme Soviet and at party conferences. Work had to be undertaken to ensure that Belarusian did not become a dead language, like Latin. Only then would Belarus 'stand as a normal nation.'[36]

Writing in the main Communist theoretical journal, the scholar A. Malashko contributed a philosophical item to the discussion on language. The language question, he remarked, had become a reality. A language constituted the link between people and creativity and national culture. He was quick to point out that concern for the native language did not imply support for national chauvinism, or opposition to 'internationalization.' The Russian language had to be retained as the language of 'intra-national' communication. Nevertheless, in recent years, there had been a reduction in the use of the Belarusian language. Belarusian-language schools had 'disappeared' in cities and raion centers. Why was this the case? First – and this statement appears to contradict his opening qualifications – there was the deepening of internationalization of public life in the country. In Belarus, because of the process of urbanization, a considerable proportion of the population, particularly young people, the workers and the intelligentsia, had concentrated in the city and found themselves in 'multi-national' collectives in which the Russian language prevailed.[37]

There were also more subjective reasons for this situation. In the 1950s and 1960s there had existed 'uncontrolled freedom' to withdraw from the study of the Belarusian language, and this phenomenon had produced the decrease in the number of Belarusian schools. The organs of popular education had allowed the problem to continue, and thus the main efforts of the party and Soviet organs were now directed to the correction of an unfortunate situation. Schools now had to be reoriented toward the study of the native language. In Russian-language kindergartens, it was necessary to increase the amount of time during which children could be acquainted with Belarusian folklore, national literature, music and national creations. Even in Russian-language schools, children must begin to learn the native language from the second grade onward. Textbooks were also anticipated, and in 1988, preparatory work had begun on a new journal entitled *Belaruskaya mova i litaratura u shkole*. A four-volume history on the working class of Belarus had recently been published, while a five-volume history of the Belarusian peasantry was in progress.[38]

One recent graduate from a Belarusian school in 1989 noted that he had recently joined the army. The recruits there came from all parts of the Soviet Union. It had come as a profound shock to him to realize that 'Ukrainians, Georgians, Lithuanians – all nationalities cherish their native language, their national values, their culture.' A man from Ukraine had addressed him directly, commenting that the Belarusian language was only spoken in the villages, and that the language was never heard in the cities. 'This resounded with reproach,' commented the graduate. He wrote to his former teacher to ignore those pupils who claimed not to need the native language. He had become ashamed of his former attitude toward the Belarusian language.[39]

Another critic of language policy, a Candidate of History, observed in *Kommunist Belorussii* that just as a man cannot have two mothers, he also cannot have two native languages. Why, he wanted to know, were 'we' spending huge sums on the study of foreign languages, especially when we the students were not fluent in them even after graduation; and why were residents of the republics all expected to learn Russian, whereas Russians living in the RSFSR were not required to learn any of the languages of the Soviet republics? For those living in Novgorod, Leningrad, Vladivostok and Moscow, it was not compulsory to have the two mothers. Why were the Russians so privileged? Other languages were clearly important in his view: English for communication outside Soviet borders; Polish for the history and culture of Belarus; Ukrainian, Latvian and Lithuanian in the border regions. But we have to start with ourselves, he added, perhaps beginning with the journal itself. It was high time that 'liberated party workers' recognized that perestroika in the national sphere was 'serious and permanent.'[40] Clearly therefore, by the late 1980s, the language question had reached a crisis point.

The question was addressed in the policies of the Belarusian Popular Front, which adopted a firm stance on the need for the primacy of the native language in the republic. On 26 January 1990, the Belarusian parliament accepted the Law 'Concerning Languages in the Belorussian SSR,' which guaranteed all citizens of the republic the right to use their native language, and made Belarusian the state language of the republic.[41] The urgency of language reform was also demonstrated in September 1990, with the adoption of a State Program on the Development of the Belarusian language and other national languages in the BSSR (it followed by only two months the declaration of state sovereignty of the republic within the Soviet Union). It noted in the preamble that the sharp reduction of the use of the Belarusian language had deprived the people of a significant part of their cultural heritage, stating that the historical and ethnic territory of the Belarusian SSR was the place of abode of the Belarusian people.[42] Belatedly, the authorities of Belarus in the late Soviet period had recognized a national crisis: that the development of the nation and its culture was irrevocably tied to the retention and usage of its national language.

That the leaders of the republic reached such a conclusion was a result largely of the efforts of the Belarusian Popular Front, which had made the language question a key component of its program. In 1993, the BPF manifesto declared its support for the state policy of developing and protecting the Belarusian language and culture, and for the priority of the Belarusian language in state and public life. It also strongly opposed the efforts to assimilate the Belarusian nation within a Russophone entity, and what it perceived as official support for bilingualism, since the latter served to reduce the impact of the Belarusian language and culture in national life. The BPF program noted that the country needed a new and modern educational system from kindergartens to universities that should be under close public supervision. This system should be devoid of ideology, and place more emphasis on the teaching of the humanities, particularly history.[43]

However, the long-term development of Belarus in the Soviet period; the consolidation of Communist forces in the parliament; and even the July 1994 presidential elections in which a run-off occurred not between a quasi-Communist and a Democrat but rather between two Communists; all suggest that the language question may not currently be a priority of the independent state.[44] Indeed one serious question is where a state might begin to redress the balance between the use of Russian and Belarusian in urban centers. One teacher of Russian has declared that the prioritizing of the Belarusian over the Russian language is correct. Both languages (in early 1989) were declared equal and served the principle of the merger of the two Slavic peoples. 'But as we live in Belarus,' the teacher stated, 'then the native language must come first of all.'[45] It seems therefore that the process of Russification can only be reversed gradually – indeed one can hardly expect a majority of city-dwellers to change languages overnight.

LANGUAGE IN THE CONTEMPORARY ERA

In demographic development and the use of the native language, Belarusians have appeared at a disadvantage compared to the Baltic states or Ukraine. The republic of Belarus is limited by its relatively small population, and by its lack of natural resources, particularly in the sphere of energy, from playing a dominant role in East European politics. Its parliament has appeared resistant to radical economic change and the introduction of market reforms. And, as we have noted, national consciousness and the progress of the native language has been limited by the way in which the republic was developed in the Soviet period. In all these respects, Belarus is at a disadvantage *vis-à-vis* its neighbors. Yet it would be misleading to portray the situation in too gloomy a light and despite a bleak economic and social picture, there are some paradoxical gains from the Soviet period.

First, Belarusian cities were notably lacking in ethnic tension in the first two years of independence. In that period there was virtually no reason other than a national census to differentiate between one Slavic nationality or another living on the territory of the republic. A sociological survey conducted by the group Public Opinion and published in 1992 provided evidence that Russians living in Belarus felt little or no commitment to Russia as a motherland. In responding to a question what they felt was their fundamental nationality, 47.5 percent of Russians surveyed declared that it was based on the territory of constant and permanent habitation. On the other hand over 64 percent of Russians denied that nationalism, i.e., of the Belarusian variety, had the potential to play a major role in society.[46] Thus though there appeared to be little likelihood of some sort of Crimean phenomenon appearing in Belarus, Russians were unlikely to lend their support to any forms of national development of the Belarusian culture. If there were to be military or economic links with Russia – and the latter seemed inevitable – then the links would be with a perceived friendly neighbor as opposed to a potential motherland.

In addition, there were some serious impediments even to the integration of the Russian and Belarusian economies, particularly on the question of a common currency. Russia was also facing uncertain political times. The Russian president had powerful enemies within the parliament and had begun to behave somewhat erratically as an international statesperson. Though some western analysts saw fit to compare Lukashenka with the extreme Russian nationalist and leader of the misnamed Liberal-Democratic Party, Vladimir Zhirinovsky, and maintained that their views were similar, this appears to have been a misapprehension. However,

Belarus remained the least *national* of the former Soviet states in 1992. The Public Opinion survey also revealed in its sampling of the Belarusian respondents that relatively few Belarusians are nationally conscious, and only a small minority supported the notion that Belarusian should be the only state language. Thus on the question of the status of languages in the Republic of Belarus, 8.9 percent of Russians surveyed and 29.3 percent of Belarusians maintained that the Belarusian language should be the state language. Almost 64 percent of Russians and a high 45.5 percent of Belarusians considered that the state languages should be Belarusian and Russian. Reflecting Mal'dis's concept of national nihilism were the responses to another question, on the understanding of the term 'motherland.' Only 35.9 percent of Belarusians and 10.9 percent of Russian respondents agreed with the statement that 'My motherland is Belarus.' Almost 10 percent of Belarusians and 19.3 percent of Russians still considered their motherland to be the former Soviet Union, while a plurality of responses for both groups favored the response that their motherland was the place in which a person was born (a somewhat meaningless answer in that it does not denote a country or state).[47]

By the spring of 1995, the question of state languages had grown more acute, prompted by the actions of the president himself. Not only was Lukashenka inept in the Belarusian language, he was ridiculed in the nationalist newspaper *Svaboda* for such failings.[48] His response, clearly backed by prominent figures of the former party hierarchy, was to demand a referendum on the status of the Russian language in Belarus and the question of its elevation to a state language. Letters to the press generally supported the adoption of Russian as a language of equal status.[49] One survey with 1100 respondents conducted in March 1995 indicated that 54 percent of those polled supported the referendum on the Russian language, 29 percent were opposed, and 18 percent did not respond.[50] Such support should not be attributed simply to attitudes of Russian chauvinism. There were often more practical bases to the responses. One academic noted that although he generally supported the advancement of the Belarusian language, the process had to take place gradually, alongside the preparation of adequate textbooks and scientific journals. At that juncture (April 1995), the Belarusian language was not yet in a position to replace Russian in all facets of life.[51] The danger, however, was that if Belarusian were to lose its position of eminence, both the language and its mother state might eventually disappear altogether.

Whether or not such responses mean that Belarus is on the road to national self-destruction as an independent state remains to be seen. Its international revival may be dependent upon its finding a role as a mediator within the CIS countries, and as a state that can forge close ties with its historical neighbors of Poland and Lithuania. The establishment of an independent state was a result of several factors, but its impetus came from the outside, with the collapse of the Union and Gorbachev's plans for a revised agreement between Moscow and the republics. It cannot be said to have occurred because of a strong movement for independence from within the country: such a movement did exist but was hardly a dominant factor.

The Soviet legacy has also provided a basis for the existence of a Republic of Belarus into the twenty-first century: a state with clearly defined borders; and a seat in the United Nations. Its history has shown on the other hand the problems that can be caused by rule from distant Moscow in terms of repressions of the past; and the threat to the very existence of the Belarusian language and national culture. These are all reasons for Belarusians today to wish to determine their own future, but international survival will depend on a greater commitment of all citizens to their state than hitherto has been exhibited.

NATIONAL CULTURE AND NUCLEAR DISASTER

What is the relationship between the dilemmas of national and cultural development and the consequences of the Chernobyl nuclear disaster in Belarus? The link is a very clear one that has rendered the response to the accident very restricted in its scope and commitment. In May 1986, despite some local efforts, there was no independent Belarusian response to Chernobyl. Relief efforts were directed from Moscow. Very seldom were republican voices heard in the immediate aftermath (that of Ales' Adamovich was a notable exception, but he was a Moscow-based Belarusian). For three years, the full import of the accident on Belarus was largely unknown outside official circles in the Russian capital. The revelations of early 1989, which indicated widespread radioactive fallout in the Homel', Mahileu and Brest regions, caused great consternation in Belarus, and within three years, the dissolution of the Soviet Union left the republic of 10.3 million alone to deal with its problems. A national emergency called for a national response; but how could there be a national response without some form of national ethos, a belief in official circles that Belarus – its declaration of independence notwithstanding – constituted a nation?

As demonstrated above, the Soviet period saw the creation of a BSSR, which contributed toward national self-awareness in that it united most ethnic Belarusian territories (with some glaring omissions) in one entity. This was not in any real sense of the word a state that existed formally. It had no power to determine its destiny and its leaders for the most part did little more than echo decisions made in Moscow. From the late 1920s onward, Soviet policy systematically destroyed any essence of a nation, partly by accident (industrialization taking precedence over other policies and resulting in the Russification of the native elite in urban centers) and partly by design (the systematic elimination of the Belarusian intelligentsia through purges and executions). The development of the republic took on a unique form in the USSR: the pace of urbanization was particularly rapid. Moreover, it resulted in the emergence of one dominant city that virtually controlled the rest of the country; a city in which the Belarusian language and culture had never been permitted to dominate. Yet it was this same city that was faced with the decisions on how to deal with Chernobyl. The future of the nation was reportedly at stake. That future was to be decided in Minsk, by politicians who were by no means all committed to the independence of their republic or its future, and who had made a habit of refusing to act without specific instructions from their superiors in Moscow.

Such a dilemma posed major problems. Yet these were exacerbated by a second phenomenon, namely that the nuclear disaster affected principally the villages. It was first and foremost a rural phenomenon that affected agricultural settlements.[52] The Belarusian identity and language existed primarily in such villages, even though the intellectual leadership was based in Minsk. Statistics cited above have demonstrated that native language retention was highest among the elderly, most often among those living in villages and hamlets, cut off from the heavily Russophone and Russified cities. The contamination of a vast area therefore was perceived by some as threatening the future of the nation not merely in health terms, but also by including in a zone of alienation important repositories of its cultural past.[53]

3 The Contaminated Zone

Almost a decade has passed since the accident of 26 April 1986 at the Chernobyl nuclear power plant in northern Ukraine. The term 'disaster' is not hyperbole given the scale involved. Assessing the economic damage to a small republic like Belarus is no easy task. In September 1989 the first estimate of the cost of the accident was provided by the republican State Planning Committee and amounted to 15 billion rubles. Six further appraisals followed at the behest of the government and Communist Party of Belarus, each of which was significantly higher than the previous one and culminating in a March 1991 figure of 129.9 billion rubles. Thereafter the question became complicated by the fluctuations in prices and currency exchange. Nonetheless, an estimate of sorts was provided by the chairman of the Supreme Soviet, Stanislau Shushkevich, who informed an international ecological conference in Rio de Janeiro that the total damage from Chernobyl to the Republic of Belarus amounted to 206 billion rubles or 16 annual republican budgets.[1] More recently, Ivan Kenik, the Minister for Extraordinary Situations and Protection of the Population from the Consequences of the Chernobyl Catastrophe, noted that according to the calculations of 'our experts,' in the period 1986–2016 the costs for dealing with the accident would amount to 32 annual budgets of the pre-accident period.[2]

The ramifications of the disaster remain mired in controversy. There is no consensus on the number of fatalities that have occurred and little agreement on the causes of a number of diseases and illnesses that have been observed within the contaminated areas. Since 1989–90, there has been a worldwide campaign to assist the victims of Chernobyl, particularly children. At the same time, some sources have argued that such aid is unnecessary since the results of Chernobyl in the medical sphere have not borne out the initial pessimistic prognoses. It is pertinent therefore to offer a reexamination of this question, and our focus is limited to the Republic of Belarus, which received the brunt of the radioactive fallout (see below), but which was the least equipped of the affected states to deal with the repercussions, both from a political and a material perspective. The intention is to answer the following questions in Chapters 3, 4 and 5:

1. What was the total radiation fallout and what concepts of radiation tolerance for the population were devised?
2. What has been the relationship between the development of Belarus as a republic and state in the Soviet period and the time of independence

and the evident inability to deal with all aspects of the Chernobyl disaster? Could one say that deficiencies in state development in this republic impeded initiative and ability to respond to a major industrial and economic catastrophe?

3. How significant is the assistance that has been given to those affected both by the government of Belarus and from nongovernment and international organizations?

4. Related to the above, how effective and expedient are the current programs whereby children from different areas of Belarus are permitted to travel outside the country for a period of rest and recuperation with international host families during the summer months?

5. How were evacuations conducted in the post-Chernobyl years and have they had a beneficial or detrimental impact upon the population of resettlers?

6. Are the resettlers suffering from psychological stress? Is it possible to provide an outline of the psychological impact of the disaster upon the population of Belarus?

7. What have been the principal medical repercussions of the Chernobyl accident in the period 1986–94?

8. Is it possible to discern a correlation between illnesses among residents of the contaminated zones, liquidators and evacuees, and increased radiation exposure as a result of the Chernobyl accident, and, if so, for which diseases in particular? Can one provide today a comprehensive picture of the health effects of Chernobyl?

9. What challenges lie ahead? Is it a viable option for the republic to pursue its own nuclear energy program given the severity of the impact of the Chernobyl nuclear accident upon its territory?

In addition to, but related to the medical information are the social and living conditions of the population, both in the contaminated zones and – in the case of the evacuees – in their new places of abode. With respect to question 6, it is difficult to underestimate the psychological stress placed on the evacuees, many of whom, for economic, social and practical reasons, have found life intolerable over the past few years. We will therefore include in the discussion the question of whether evacuations were advisable, the levels of radiation tolerance to which the population was subjected, and how far it is possible to include the evacuees and residents of the contaminated zone within the category of victims of Chernobyl. This problem also encompasses the matter of state and nongovernmental aid, its effectiveness and current results.

RADIATION FALLOUT

The accident released over 50 million curies of radioactive material, the fallout from which encompassed a population of 17 million people in the former Soviet Union, including 2.5 million children under the age of five. About 70 percent of the fallout landed on the territory of Belarus (then the Byelorussian SSR).[3] Observing only the fallout of cesium-137, the most prevalent radioisotopes in the soil today, an estimated 40 400 square kilometers of Belarusian territory were contaminated with more than 1 curie per square kilometer (1 cu/km^2), or 19.5 percent of the total area of the republic, with a population of 1.8–2.2 million people.[4] The number of children in this group is cited as 440 000, among whom in 1991 on territories contaminated by 15–40 curies of cesium per square kilometer in the soil were to be found 31 000; in the region of 5–15 curies, 41 000; and in the area of 1–5 curies, 365 000.[5] Though the figures and determination of areas to be evacuated pertain to levels of cesium, it should be emphasized that the fallout of radioactive iodine encompassed about 80 percent of the republic – only the most northerly regions avoided fallout – and that levels in many places exceeded 1000 curies per square kilometer.[6] Because the half-life of iodine-131 is so short, at 8 days, this radioisotope is usually omitted from long term prognostications and analysis of the results of contamination. However, as will be demonstrated, to date it is iodine-131 that has provided the most harmful medical consequences.

The areas most contaminated with cesium were the oblasts of Homel' and Mahileu, though more specifically, four major areas of fallout have been designated:

1. A central region to the west and northwest of the city of Minsk, which contains isolated patches of cesium at levels in the soil of 1–3 cu/km^2, and an occasional spot with 5 cu/km^2 and above. It encompasses parts of 10 raions: Valozhyn, Barysau, Byarezina, Salihorsk, Maladechna, Vileyka, Stoubtsy, Krupki, Lahoisk and Slutsk. The westernmost settlement is that of Byarozauka, located at a central point between the cities of Minsk and Hrodna.[7]

2. A region in the southwest that embraces the southern part of the Pripyat Polissya as far north as the city of Pinsk, and where levels of cesium reach as high as 15 cu/km^2, but where a figure of 3 cu/km^2 is more prevalent. The most westerly point is the town of Drahichyn, located about 100 kilometers east of Brest.[8] Six raions of Brest Oblast have been partially contaminated: Luninets, Stolin, Pinsk, Drahichyn,

Byaroza and Baranavichy, in addition to six raions of Hrodna Oblast: Dyatlava, Iue, Karelichy, Lida, Navahrudak and Smarhon.

3. An eastern region that covers the eastern districts of Mahileu Oblast, with levels of 15–40 curies and above. In 1987, it was discovered (though not revealed publicly for two years) that there were some areas of very high cesium contamination in the raions of Novoel'ne, Krasnapolle, Chudni, Malinovke and Cherykau, in which levels were recorded over 100 cu/km^2; in addition to seven other regions with over 80 cu/km^2.[9] There is also a substantial area around the city of Chavusy in which levels vary from 1 to 10 cu/km^2.

4. The southeastern region that includes the south and southeastern districts of Homel' Oblast, with cesium levels around 40 cu/km^2, in addition to significant fallout of strontium-90. In the southern parts of the Khoiniki and Brahin raions (parts of which are included in the zone evacuated in 1986), isotopes of plutonium-239 and plutonium-240, with levels in the soil up to 0.1 cu/km^2 have also been ascertained.[10]

In addition to these four main areas, there were also found four settlements in Vitsebsk Oblast with cesium-137 levels in the soil of more than 1 cu/km^2. Among the affected settlements are reportedly some major population centers. They include the republic's second largest city, Homel' (population 500 000); the southern outskirts of the city of Mahileu (356 000); and areas of Mazyr (101 000), Rechytsa (69 000), and smaller towns such as Dobrush, Kalinkavichy, and Naroulya.[11]

One source puts into perspective the proportion of total fallout from Chernobyl irradiation in Belarus by noting that 59 percent of the fallout of cesium in the soil of over 5 cu/km^2 lies in the republic; 59.3 percent of the 15–40 cu/km^2 area; and 69.4 percent of that over 40 cu/km^2.[12] The other areas of significant fallout were the northern regions of Ukraine (particularly the Kyiv, Chernihiv and Zhytomyr oblasts);[13] and the Bryansk Oblast of Russia.[14] It has also been observed that Belarus possesses 400 settlements with over 15 cu/km^2 of cesium in the soil, compared to 206 in Russia, and 49 in Ukraine; in addition to 48 with 40–60 cu/km^2; and 22 with over 60 curies, as opposed to only eight in Russia and none in Ukraine. This same source states that though strontium-90 was confined largely to the 30-kilometer zone, density levels of over 5 cu/km^2 can be found well outside the initial zone of evacuation. Further plutonium fallout reached as far as the eastern border of the republic.[15] Additional information on the total fallout from Chernobyl is found in Table 3.1.

Table 3.1   *Radiation fallout from Chernobyl*

Level	Republic	Total area (sq. km)
5–15 curies	Belarus	10 160
	Russia	5 760
	Ukraine	1 960
14–40 curies	Belarus	4 210
	Russia	2 060
	Ukraine	820
Over 40 curies	Belarus	2 150
	Russia	310
	Ukraine	640[16]

Source: *Gomel'skaya pravda*, 8 June 1993.

Perhaps the most important characteristic of the radiation fallout is its lack of uniformity. It is possible to find areas at considerable distance from the destroyed reactor where levels are significantly higher than in the reactor vicinity. Some areas in which contamination of the soil by cesium averages 15–40 cu/km^2 have not brought about high internal doses of the population in the vicinity, whereas some areas with only 1–5 curies appear to have had a significant impact on the local residents.[17] Much depends also on the nature of the soil. In sandy soils, particles tend to be found in the upper layers mainly at a level around 5 centimeters (cm) from the surface, whereas in the peat-boggy soils that tend to predominate in southern Belarus, they can be found at much lower levels down to 30 cm. Strontium-90 migration along the food chain has been found to be more intensive than that of cesium-137,[18] even though the amounts are much lower. Finally, the most pervasive of all radioisotopes was the short-lived iodine-131, the fallout of which encompassed all the republic with the exception of the northernmost Vitsebsk Oblast,[19] but which has a half-life of only 8 days, compared to the 29 years for strontium-90 and 30 years for cesium-137. Nonetheless, as will be demonstrated, it is iodine that has had the most immediate and clearly discernible health effects upon the population.

EVACUATIONS AND STATE ASSISTANCE

In 1986, an estimated 116 000 people were evacuated from the zone of 30-kilometer radius around the destroyed reactor, of which 24 700 were from the Belarusian side of the border. These residents were moved from 107 settlements of the Brahin, Naroulya and Khoiniki raions of the Homel' Oblast. It has been calculated that a further 652 000 Belarusian residents were left behind in areas contaminated by radionuclides in which it was necessary to introduce restrictions on the use of locally produced milk because of the high content of radionuclides therein. By June 1986, the Belarusian Academy of Sciences and the State Hydrometeorological Committee had advised the republican government of the situation in Mahileu Oblast. Particular concern was expressed for 50 settlements in the raions of Cherykau, Slauharad, Krasnopolle, Kastsyukovichy and Bykhau raions. Despite such warnings, no evacuations took place until January 1990 and then they encompassed only a portion of the towns and villages affected, and reportedly were carried out 'at an extremely slow pace.'[20] One can deduce from the above that the health of residents may have been imperiled by the slow and half-hearted reaction of the all-Union and republican authorities.

Belarusian state reaction to the Chernobyl disaster can be divided into several stages. In the period from 1986 to the spring of 1989, official secrecy in Moscow precluded a coordinated program. Information on health effects was officially classified and the task of dealing with Chernobyl was in the hands of a government commission at the all-Union level. Without a detailed knowledge of the precise areas of fallout, the republican administrations of Belarus and Ukraine could not formulate a program designating which areas should be evacuated and which could survive at certain levels of radiation. Between 1989 and 1991, greater openness at the all-Union level permitted the enactment of a number of laws pertaining to Chernobyl. In July 1989, the XI Session of the Belarusian Supreme Soviet held the first discussion of the proposed State Program for the Liquidation of the Consequences of the Chernobyl Disaster, which was approved in final form at the following session in October 1989. It established habitation restrictions and ordered the evacuation of over 118 000 people living in 527 settlements with cesium levels in the soil of over 15 cu/km^2.[21] At this time a permanent commission on Chernobyl was established within the Supreme Soviet.[22]

The 1989 Program also noted the intention of the government to provide the population with food according to medical norms in all areas in which the density of contamination with radioactive cesium exceeded 1 curie per square kilometer. This figure thereafter stood as the minimum level that warranted official concern and consequently the territory encompassed by the effects of Chernobyl became much wider than hitherto, i.e., it was extended beyond the Homel' and Mahileu regions. From the start of

1990, according to the State Program, children in kindergartens and schools in the affected regions were entitled to free supplies of clean food. Financial benefits for the population at large were also introduced. The pensionable age of men was lowered to 55 years and that of women to 50 years, with the extension of contiguous vacations for those living in the zone to 30 calendar days. One-time stipends were issued for the birth of children. Poor families could receive a grant of 24 rubles per month (at that time about 10 percent of the average monthly salary) for each child under the age of 14, with the total income not to exceed 75 rubles per month through this means. Differential levels of payment were introduced that were dependent upon the level of contamination in the soil, from an increase of 30 percent of the average salary in the realm of 5–15 curies to 100 percent at over 40 curies. A further payment of 30 rubles per month was offered for every family member living in settlements with restricted use of food products. Finally those resettled were to be recompensed for the cost prices of construction in their new abodes (residences, gardens, *dachas*, garages, and economic buildings).[23]

The State Program was ambitious and well intentioned but it was never more than partially fulfilled in any given year. The logistics involved in the transferral of people and designation of territories were overwhelming. Moreover, the Program's course was overtaken by political events unconnected with Chernobyl which resulted in the sudden withdrawal of Union funds assigned to the program, and was barely compensated at all during the reassignment of Chernobyl funds following the dissolution of the Soviet Union (Belarus received about 4 percent of these funds). Thus it was left to the emergent state of Belarus at the end of the Soviet and beginning of the post-Soviet period to commit new state funds to assist those suffering from the impact of the Chernobyl disaster.

THE LAWS OF 1991

In 1991, the Supreme Soviet of Belarus accepted two new Chernobyl laws. The first, enacted on 22 February (with some additions on 11 December 1991), was entitled 'Concerning the Social Protection of Citizens Suffering from the Catastrophe at the Chernobyl Nuclear Power Plant.' It affixed legally the rights of the population affected by radiation fallout, particularly the liquidators, evacuees and citizens living in the vicinity who had moved voluntarily to new places of residence. It established a radiation tolerance level for the population (an effective equivalent dose of irradiation) at no more than 0.1 rems per year (see below).

Those who had suffered from the accident – both 'liquidators and victims' – were entitled under this law to free medical service, annual recuperation without cost at sanatoria, and reductions of 50 percent in rental payments and the costs of utilities at their residences. The law also affixed significant financial compensation to those living in contaminated zones that were not subject to evacuation, at 105–210 and 395–1415 rubles according to the degree of contamination of the soil. The victims of Chernobyl also had the right to an early pension.[24]

On 12 November 1991 the Republic of Belarus accepted a 'Law concerning the Correct Regimen of the Territory Suffering from Radioactive Contamination as a Result of the Accident at Chernobyl Atomic Energy Station,' which regulated the conditions of habitation, and economic and scientific-research activities in the contaminated zone. The law defined the contaminated territory cited in the previous law as consisting of five zones, as follows:

1. The Zone of Alienation, which consisted of the territory around the reactor and which was evacuated in the first weeks after the accident, commencing 27 April 1986. In addition to the Belarusian territory in the designated zone of 30-kilometers' radius around the reactor, it also included an additional zone of evacuation of areas in which the density of radionuclides in the soil exceeded more than 3 cu/km^2 for strontium-90 and over 0.1 cu/km^2 for plutonium-238, 239, 240, and 241. In this zone, the only permissible activity was that linked to the provision of radiation safety, such as the removal of radioactive substances, scientific research and experimental work.

2. The Zone of Primary Evacuation, an area in which contamination levels of the soil with cesium-137 exceeded 40 cu/km^2; for strontium-90, levels of over 3 cu/km^2; and for plutonium, over 0.1 cu/km^2.

3. The Zone of Subsequent Evacuation, a territory with contamination levels for cesium-137 of 15–40 cu/km^2; for strontium-90, 2–3 cu/km^2; and for plutonium-238, 239, 240 and 241, from 0.05 to 0.1 cu/km^2, in which the annual average equivalent dose (AAED) of irradiation per person may exceed 0.5 rems per year, and other areas with lower density of radionuclide contamination in which the AAED would exceed that figure.

4. The Zone with the Right to Evacuation; an area with cesium-137 levels of 5–15 cu/km^2 or with strontium-90 levels of 0.5–2 cu/km^2; or plutonium levels of 0.02–0.05 cu/km^2 in which the AAED may exceed 0.1 rems per year over the natural and technogenic background.

5. The Zone of Periodic Radiation Control; the territory with cesium-137 levels of 1–5 cu/km^2; strontium-90 levels of 0.15–0.5 cu/km^2; or plutonium levels of 0.01–0.02 cu/km^2, in which the AAED might exceed an additional 0.1 rems per person per year.[25]

Administration over the various zones was placed in the hands of the Council of Ministers of the republic, local soviets and the State Chernobyl Committee,[26] and responsibility for assessing the radiation system was in the hands of the Belarusian Republican Administration for Hydrometeorology. Each population point in irradiated zones had to be issued with a radiation-ecological passport. In addition, the State Chernobyl Committee was ordered to create local centers of radiation control affiliated to the local Soviets, with responsibility for ensuring that citizens were provided with food products that conformed to radiation standards.[27]

In between the two laws the Belarusian authorities approved the funding of the State Program for 1990–5 to deal with the effects of Chernobyl.[28] The budget assigned for the 1990–5 program to overcome the consequences of Chernobyl was 49.8 billion rubles; with a further 37 billion rubles anticipated for the period 1996–2000. It allowed for the evacuation of residents from 112 population points in which each individual would receive an equivalent irradiation dose of more than 0.5 rems per year, in addition to population points with cesium levels in the soil between 15 and 40 curies, and families with children under the age of 14 who lived in areas in which agricultural activity was forbidden. It also stipulated that in the areas of constant and periodic control appropriate medical centers be established with laboratories for the diagnosis of oncological illnesses.

By 1991, new limits for radiation tolerance were in effect consisting of 1 millisievert (mSv) per annum additional radiation per person (0.1 rems), compared to previous levels that had been established by the former USSR National Commission for Radiological Protection in 1988 at 5 mSv (0.5 rems), based on an 'acceptable' lifetime dose of 350 mSv (35 rems). The above law was to be implemented from 1 January 1990: thus it remained in place for only one year.[29] The lower level of tolerance followed extensive research and a concept for habitation developed by the Belarusian Academy of Sciences. It stipulated that the following criteria had to be taken into consideration before permitting habitation: the amount of radioactive fallout in the territory; the dose load received by the population in addition to the natural background; and the availability and effectiveness of protective measures to reduce the radioactive impact.[30] Though comprehensive, this program proved to be well-nigh unworkable, not least because of the broad area now encompassed in the evacuation zones. After declaring independence in the summer of 1991 and with the final dissolution of the Soviet Union in December of that year, the republic found itself facing an economic crisis, with few funds available to deal with a disaster of the magnitude of Chernobyl. By 1993, it was evident that programs for dealing with several aspects of the disaster at the state level had been virtually abandoned.[31]

In the summer of 1993, the government set up a commission led by radiochemist and doctor Yaugen Pyatraeu (Yevgeniy Petrayev), which was given the task of making 'corrections' to the 1991 conceptions for living in contaminated territory. The commission was to adhere to the original stipulation that individual exposure should not exceed 1 mSv per year; that protective measures would be necessary at exposure levels of 1–5 mSv; and that at more than 5mSv, the population must either be evacuated or levels lowered to an acceptable level. The Pyatraeu Commission's purpose was to change the nature of the original law so that it took into account more accurately the quantitative effect of ionizing radiation on humans, i.e., it would use as a basis for Chernobyl laws the dose load received by the population rather than the radiation levels in the soil.[32] However, in reality, it seems that the goal of the commission was to reduce significantly the area of dangerous contamination by somewhat artificial means, since costs of dealing with the tragedy's effects were well beyond the means of the Republic of Belarus.[33]

A critique of the new concept was offered at the draft stage in August 1993 by Valeriy Shumilau, a Candidate of Medical Sciences who played a part in the original clean-up campaign in the immediate aftermath of Chernobyl. Shumilau pointed out that the Pyatraeu concept would permit one to move to a single criterion for habitation of contaminated areas, namely the annual average equivalent dose (AAED) received by the population. If such proposals were accepted, he added, then almost all regions with cesium levels in the soil of 1–5 cu/km^2 would lose their social and radiation protection. He noted also that the data on radiation exposure were far from precise, and that the new concept tended to ignore the actual living conditions of the population. Though ostensibly the government had allocated ample funds to deal with the effects of Chernobyl, most of these had been put into decontamination campaigns and the construction of new townships. Two of the original three yardsticks for protecting the population – contamination of the area and the effectiveness of protection measures – had been eliminated by the new concept. Only the AAED remained.[34] By September, however, the new concept had been approved by the Supreme Soviet Committee devoted to problems of Chernobyl; by

Belarus

the Belarusian Academy of Sciences; the National Committee on Radiation Safety; and the State Chernobyl Committee.[35] Nonetheless, it was acknowledged that even with a revised concept, the government had sufficient funds to deal with only 58.4 percent of the total costs arising.[36]

A poignant illustration of the woeful condition of state assistance for the effects of Chernobyl and the gap between what was to be provided on paper and the reality was produced in late 1992 by attacks on the government. Two of these are notable: the first was written by V. Chirun and L. Zverev, and entitled 'How Healthy is the Ministry of Health?', while the second was an open letter to Vyachaslau Kebich, Premier of Belarus, from two parliamentary deputies, L. Zverev and G. Grushevoy, under the headline: 'Will the National Genetic Program be Fulfilled?' Together they represent a serious indictment of the authorities, and in particular of their failure to resolve financial predicaments that involve health risks to the population at large.

Chirun held the position of head of the military-medical service in the Homel' committee of state security, while Zverev held a number of positions, including that of a people's deputy, member of the State Chernobyl Committee, and the status of a qualified doctor. They pointed out that medical care for the population of Homel' and Mahileu oblasts in 1991 and 1992 depended on work brigades that had been dispatched into the contaminated regions. Reportedly 82 such brigades were operative in 1991. While there were few problems with regard to the number of people working in these areas, the quality of care left much to be desired. The principal problem, in the authors' view, had been the failure to promote doctors or medical specialists as members of the brigades. At medical-prophylactic points in Homel' region, only 56.2 percent of those on duty had medical training, while in Mahileu region the figure was 57.5 percent. This lack of expertise was said to be particularly noticeable in Brahin and Krasnapolle raions where the situation was the most serious.[37]

As new doctors were being trained, they were not being made aware of the characteristic features of the ecological situation in the republic.[38] Although over the previous two years a new hospital and a polyclinic for outpatients had been built, construction work on new medical facilities was well behind schedule. Moreover, the allotted budget for such enterprises showed a shortfall of some 12 million rubles.[39] As far as aid to those raions suffering the most from the Chernobyl disaster, there was an appreciable difference between that offered to families from the zones of evacuation and those who lived in the contaminated regions. In the latter case, it was noted, official aid had been woefully insufficient and had suffered from a 'stingy budget.' This applied especially to the provision of convertible rubles that could be used to purchase medical equipment and facilities outside the borders of Belarus.

In 1990, the government allocated its funds for dealing with the consequences of Chernobyl for the period 1991–5. The cited figures for convertible rubles in 1991 and 1992 were 302.9 million and 302.6 million respectively. The actual amounts expended in these years, however, were 25 and 31 million rubles, i.e., only 8 percent and 10 percent of the originally designated total respectively. In this instance, part of the problem was the shrinking of the republican budget as a result of the declaration of independence by Belarus and the subsequent collapse of the USSR. This same factor restricted the supply of drugs to pharmaceutical outlets in the republic. These outlets were reputedly spending 10–20 times more money on the acquisition of such drugs than they could recoup by selling such products in the stores. Thus the state was being asked to supply about one billion rubles to compensate pharmacists for their indebtedness. In addition, the medical supplies on hand represented just over 50 percent of the amount in stock one year earlier.

Finally, Zverev and Chirun turned to the subject of care for those suffering from Chernobyl and the Law on the 'social protection of citizens' issued formally in the spring of 1992 by the government. The official registry of citizens suffering from Chernobyl listed only 190 000 names (compared to the 2.2 million people reported to be living in affected regions cited above).[40] The new law was reportedly being put into practice 'extremely sluggishly' despite the fact that its application had been deemed as urgent by the State Chernobyl Committee. There was, in the authors' view, too much talking by the authorities, and not enough action.

Zverev and Grushevoy published their article as an 'Open Letter' to the Chairman of the Belarusian Council of Ministers, Vyachaslau F. Kebich. Grushevoy is the Chairman of the Belarusian Charitable Fund 'For the Children of Chernobyl' (see below) and at that time a people's deputy for the Belarusian Popular Front (he later became an independent deputy). He made plain his disaffection with the government of Shuskevich and Kebich, and has little faith in the Lukashenka administration.[41] He regards the actions of the Belarusian government in dealing with the consequences of Chernobyl as obstructive to the efforts of his own organization to assist children in particular.[42] The letter, authored jointly with Zverev, reflects this attitude. If Mr. Kebich, the authors assumed, supported the 'National Program to deal with the Genetic Consequences of Chernobyl,'[43] then 'it is impossible to comprehend' why with such 'exceptional obstinacy' the government had failed to fulfill the program's mandates. In 1992, the authors pointed out, the frequency of congenital defects in newborns in the republic had risen by 40 percent compared with the pre-accident period, and such defects in 1992 constituted the chief cause of infant mortality.[44]

The key difficulty once again appeared to be that of financing the program. Its goal was to reduce the number of abnormalities in newborns by 20 percent, to cut infant mortality by 25 percent, and to reduce the number of child invalids by 30 percent. However, in 1991, the Program lacked both hard currency and ruble financing. Belatedly, at the end of 1991, the Ministry of Health had allotted 3 million rubles to the program. A similar story was related for 1992, with a total absence of hard currency funding, which resulted in a lack of provision of screening equipment to monitor the various congenital sicknesses among newborns. After 20 December 1992, the authors noted, the republic would abandon the bio-chemical screening of pregnant women. Thus a combination of the failure to purchase modern equipment and the cessation of even the most basic screening of risk groups had presented a moral and health crisis, in the face of which the authorities appeared to be paralyzed.

Zverev and Grushevoy argued that money was not the key difficulty because the program, if implemented fully, would result in a reduction of expenditure on health care by at least 10 times. In their view, one had to take into account the cost of subsidies for families with invalid children, the maintenance of homes for invalids, and nursing and support for patients with incurable diseases. In the meantime, they remarked scornfully, the government had managed to discover hard currency resources to finance cross-country races, and the production of tennis rackets, video equipment and automobiles. Moreover, the sale of Belarusian-made cars, tractors and chemicals would be sufficient to obtain the $700 000–$800 000 per year – which, they declared, represented about 8¢ US per head of population – required to provide up-to-date medical-genetic care for the entire population of the republic.

Materials published in 1994 by the State Chernobyl Committee make plain the widespread dissatisfaction with state efforts to alleviate problems by the population at large. It was reported to be unhappy with all aspects of life and labor; to have neither confidence nor belief in the Committee's ability to overcome the consequences of radiation contamination and the concomitant social upheaval. The chief concern of the public is said to be the health question, while almost one-third of young parents were said to be suffering from 'radiophobia,' i.e., fear of radiation, and particularly its effects on their children. Only 2 percent of those surveyed were happy with the measures of 'social protection' introduced by state organs, and there appears to be general disillusionment with trade unions and other organizations.[45]

FURTHER EVACUATIONS OF THE POPULATION

Evacuation has become an emotional issue and the subject of intense debates. It has been argued, for example, that to move people from their native places of residence is, *ipso facto*, to endanger their health by raising stress levels and causing psychological anxiety:

> It appears that due account has not been taken by the authorities of the many negative aspects of relocation in formulating relocation policy. There are indications from studies in other areas that the mass relocation of people leads to a reduction in average life expectancy (through increased stress and changes of lifestyle) and a reduced quality of life in a new habitat.[46]

There is also evidence to suggest that attachment to one's native place of residence was much greater in the areas affected by Chernobyl than in most other parts of Europe. Further, such attachment pertained especially to the elderly, who made up a considerable portion of the population in such regions. To where would populations be moved, and would they be moved by settlement or as individual families? All the governments of irradiated territories faced such questions. As is well known, the Belarusian territories composed the northern, smaller portion of the original 30-kilometer zone around the Chernobyl reactor, which was evacuated in April–June 1986. On the Belarusian side of the border, some 24 600 inhabitants were initially moved from their homes.[47] These people were residents primarily of the Brahin and Khoiniki raions of the Homel' Oblast. Subsequently another 50 000 inhabitants were evacuated from this same oblast from 13 805 families.[48]

There are indications, however, that the authorities initially were reluctant to begin new evacuations after those from the 30-kilometer zone in the summer of 1986. Thus in Mahileu Oblast, heavily contaminated zones were discovered in 1987 and it was stipulated that 1494 people were to be evacuated. This order had not been carried out by 1990, ostensibly because Boris Shcherbina, the Chairman of the USSR Government Commission on Chernobyl, had visited certain farms – including one in Cherykau Raion in 1987 – and informed the population that there was no cause for concern.[49] In the period 1986–91, it can be asserted that generally the population living in zones with contamination levels of more than 5 curies/km^2 of cesium in the soil was anxious to be moved. Of 2117 families who lived in Brahin early in 1991, for example, it was stated that only 78 had expressed a desire to remain in their home.[50] The evacuations had already fallen well behind schedule.

In the period 1991–3, according to one source, 13 900 families composed of 36 700 persons were evacuated from contaminated regions of Belarus, yet this figure constituted less than half of the numbers designated for removal at this time. Of this number, 10 500 families were from Homel' Oblast (28 500 people) and 3400 from Mahileu (8000 people). The basic problem was that there were no residences available for many of the resettlers.[51] Academician E.F. Kanaplya, one of the key figures involved in the scientific analysis of the impact of Chernobyl in the republic, noted in the spring of 1993 that overall, in the period 1986–93, approximately 70 000 people had been evacuated out of a designated total of 100 000. In his view a major problem was the dispersal of evacuees to different settlements, which was having 'an adverse psychological effect' on them.[52] Higher figures were provided at this same time by another source, which indicated that over a seven-year period, over 125 000 people had been removed from the contaminated areas.[53]

Alternative figures provided by a Minsk source and encompassing the years 1991–3 suggest that the evacuations were conducted at well below 50 percent of the designated schedule (Table 3.2). In the period 1991–2 they indicate that 32 882 families were on the list for obligatory resettlement, a total of 83 330 individuals. However, only 20 893 families with 57 043 people were actually moved. In the period January to June 1993 in Homel' Oblast, the plan called for the removal of 6915 families (39 761 persons), but by the end of this period, only 869 (2254 people) had been resettled, or a mere 5.7 percent of the plan. Major problems had been encountered in Vetka Raion, for which had been anticipated the evacuation of 10 390 individuals to noncontaminated zones. In practice, only 298 were moved. In this six-month period in Homel' Oblast, the Naroulya region saw the largest number of completed evacuations, but even here the total number of people moved, at 886, was only 9.3 percent of the

Table 3.2   *Evacuations conducted in 1991–92, by oblast*

Oblast	No. of families	No. of people
Brest	289	923
Homel'	16 001	44 301
Mahileu	4 603	11 819
Total	20 893	57 043

Source: Figures provided by Sergey Laptev, Minsk City Council, 1994.

Table 3.3   *Evacuations conducted January–June 1993, by raion*

	Families	People	Percent of plan
Homel' Oblast	869	2254	5.7
Brahin Raion	128	368	12.9
Buda-Kashaleuski	1	2	0.5
Vetka	131	298	2.9
Dobrush	29	57	5.4
El'sk	0	0[55]	0
Karma	140	378	6.5
Lel'chytsi	2	4	6.1
Naroulya	328	886	9.3
Khoiniki	58	112	4.1
Chachersk	52	149	2.2
Mahileu Oblast	114	246	9.6
Klimavichy	0	0[56]	0
Kastsyukovichy	23	45	10.4
Slauharad	4	13	2.0
Cherykau	9	18	7.3

Source: Ibid.

planned total. In Mahileu Oblast, the situation was even worse. Of a scheduled evacuation total of 1297 families with 2571 individuals for the first six months of 1993, only 114 families with 246 people were actually resettled.[54] One can deduce from these statistics that the state had failed woefully to fulfill its mandate in the period 1991–3.

A considerable number of those designated for resettlement had chosen to remain behind. In 22 townships, some elderly people were living in zones in which the annual dose of radiation exceeded 5 mSv, or 5 times the maximum permissible level. They were refusing to leave their homes. By 1993, numerous apartments built for resettlers remained vacant. Though many people had been declared eligible for evacuation as Chernobyl victims, almost 2000 apartments in the 'clean zone' of the republic remained empty.[57] Arguably then, elderly people were endangering their health by stubbornly refusing to leave their native villages. There appears to have been a change in attitude in the period 1991–3, from a desire to be moved at all costs, to some or even extreme reluctance to be moved. Some 432 apartments built for evacuees in Homel' oblast remained vacant in the summer of 1993. Inhabitants of the Chachersk area refused to move to 1152 apartments constructed in the oblasts of Brest, Vitsebsk and Hrodna.[58]

The true picture was more complex than a simple unwillingness to leave one's home. In July 1994, Alyaksandr Lukashenka was elected the first president of Belarus largely on the basis of his previous work as chairman of a government commission to investigate corruption. Indeed his overwhelming victory over his rival Vyachaslau Kebich may be attributed at least in part to public concern for increased corruption in every walk of life. Such corruption – though possibly exaggerated for political gains – was also manifested in the State Chernobyl Committee, and also in the case of Chernobyl resettlement. It was noted that the housing program for resettlers was a failure in several areas, including Rahachou, Dobrush and Homel'; and that in 1992, the costs of such programs had been inflated on 32 projects by some 4.56 million rubles; and in 1993 on 18 projects and 18.4 million rubles. Yet no charges had been laid and the lawbreakers were continuing such practices freely.[59]

In 1991–2, the local soviets of Belarus were obligated to assist 26 786 families to move from the contaminated territories, but less than half of this number had received help. Some 1032 apartments had been given to local inhabitants who had no rights to them. In some cases, people from outside the contaminated areas were occupying the new apartments in addition to retaining their own. Seventeen leaders of the local councils in Karma Raion had thus obtained apartments in the city of Minsk, but retained their original homes. Similar cases had arisen in the regions of Vetka, Naroulya and Chachersk.[60] In another case, the housing constructed for resettlers in the village of Khodarauka by builders from Brest was inadequate and not accepted by the local housing commission. Rather than improve their edifices, the constructors simply abandoned their work, leaving the houses free for looters to steal from them.[61]

Such cases are just one reflection of the sort of trauma endured by those assigned the status of evacuees. Evidence suggests that from the very outset of the evacuation and reconstruction program in 1986, and particularly after the collapse of the USSR in 1991, the republican authorities were inept and hopelessly unprepared for such a task. The evidence for such a statement is substantial. First, the new settlements were often selected without the necessary thorough examination of the local environment. In Homel' Oblast, for example, 26 new settlements were constructed in an area that was subsequently revealed to have radiation levels of cesium in the soil of 5–15 curies/km^2, while at the new settlement of Mayskiy, in Mahileu Oblast, objects of a social-cultural nature were built in areas with high levels of radioactive cesium.[62]

A poignant story concerns a large family from Mahileu Oblast. Over a period of five years, they had built a new house. After the Chernobyl disaster, they were ordered to move and departed for Ukraine, where their daughter lived. With their compensation money, they purchased a house in one of the more affluent areas of Ukraine. Their new neighbors, however, were reportedly far from welcoming and accused them of spreading radiation, though the article alleges that the real reason for such hostility was their relative poverty. Eventually they returned to Belarus. On Remembrance Day, they visited their former village in Mahileu, sat down in their old home and wept together. Such stories, it was added, are typical of thousands and thousands of evacuees.[63]

For the most part, the resettlers appear to have been reluctant to change their surroundings for a very uncertain future. According to one account, the new villages always seemed to be squeezed into 'tight corners,' far from hospitals, schools, stores, and suitable places of employment. No infrastructure had been constructed alongside the new houses. In some areas (Razantsy, Mishi, Matsynty), the houses had evidently been put up too rapidly. The walls 'crept' and the heating elements often gave off no heat.[64] Later in 1993, the settlement of Tupichnya appeared on the map in the Kastsyukovichy Raion of Mahileu Oblast. The houses there were reported to have cracked walls, holes in floors, flooded basements, broken-down ovens, and nonfunctioning drains. The heating plant, bathhouse and kindergarten had not been completed, while the store and school existed only as blueprint plans. And Tupichnya was far from unique among the new settlements of Mahileu Oblast. Who was to blame for such fiascos? The Regional Executive Committees of the local councils, and in particular the deputy chairmen of the same, had accepted the incomplete houses and buildings. In addition, because of inflation, it was not profitable for the building companies to work efficiently. Companies were not penalized for shoddy work and the State Chernobyl Committee's representative in Mahileu Oblast declared that huge losses had been incurred and could not be recouped on Chernobyl construction projects.[65]

Perhaps the situation was worse in Mahileu than elsewhere. In 1994, the deputy chairman of Mahileu Oblast Executive Committee, Aleksey Semkin, stated that some 90 percent of those eligible for evacuation had moved, but 1178 families remained, including 123 with children. Some were evidently unwilling to be resettled because of their attachment to their native villages. The implication, however, was that such feelings applied to a minority. Most were more concerned about the sort of future that could be expected in the new village settlements, particularly in the difficult social and economic situation predominating in the country. The settlements lacked schools, kindergartens, stores, canteens and hospitals. Potential settlers wished to be informed about the supply of gas, water, electricity, and heat. No jobs were available in the new abode. In some settlements, residents were sleeping in outdoor clothes for the second successive winter because of irregular supplies of fuel to the village boilers. The lack of hot water made it practically impossible to shower or to wash clothes.[66]

Though Mahileu may have been a worst-case scenario, exacerbated by its proximity to the contaminated zones, it was far from unique. The relatively distant Vitsebsk Oblast, for example, suffered from a similar predicament. In 1990, the oblast had received 2388 families, with a total of 8793 people, including 388 children from a primary relocation zone around the nuclear reactor. But many districts and zones of the oblast had not fulfilled their resettlement plans because the constructed dwellings were simply unfit for habitation. Consequently, the Belarusian Council of Ministers demanded that the oblast should receive a further 1315 families in 1991 of which 511 were to be from rural settlements and 804 from urban centers.[67] Not only were such problems of concern to the Belarusian authorities, they constituted a major source of anxiety among the evacuees and potential evacuees. As it came to be realized that new housing conditions were so unsatisfactory, an increasing number of families either could not be moved or refused to move. Many tried to seek removal to major cities rather than villages: 157 families had applied to be moved to Minsk; 34 to the city of Mahileu; and 19 to Babruisk. Some 409 families had asked to be moved to oblast and raion centers.[68]

In late July 1994, I.A. Kenik, chairman of the State Chernobyl Committee, was questioned about the state of the evacuation process. He declared that 'mass relocation' of citizens had been completed. Applications for future resettlement from those in the zone would be treated on the basis of removal from a town to a town, and a village to a village. He did not believe that a significant number of resettlers would be unable to find employment. First of all, he pointed out, there were 52 770 registered unemployed in the republic, yet only 775 of these were resettlers. Local soviets were under orders to give the evacuees first priority when assigning jobs and the 'Law on the Social Protection of Citizens affected by the Chernobyl Catastrophe' also had guaranteed certain rights and privileges to victims of Chernobyl. Article 22 of this law even ensured, in theory, that those who had left private homes in the contaminated zone would receive new houses or apartments as private property, even if the apartment building was state owned.[69]

Kenik clearly intended to provide reassurances to a nervous public. His comments were far from convincing and indeed are not supported by available evidence, which indicated that a large number of resettlers had been unable to find work in their new places of abode. The figure for unemployed in the republic was also unreliable since it excluded the 'hidden unemployed': those who were not officially registered and others who were either not seeking jobs or working only part-time. One of the chief problems for evacuees has been the problem of obtaining equivalent positions after resettlement. Belarus, as noted earlier, is a republic in which the development of heavy industry has been confined to a few centers, and in which the main urban areas are already heavily overpopulated. The evacuation process had thus been greatly hindered by the demographic development of the state.

Paradoxically, the virtual completion of the state program for evacuation by late 1993 made the situation even worse, as a cash-strapped government decided that capital investment in housing construction could now be drastically reduced (by 2.1 times) in 1994.[70] It would also be simplistic to assume that all the problems for evacuees lay in the inadequacies of village settlements. An exodus to the capital city of Minsk occurred after the disaster, and evidently reached mass proportions in the period between September 1993 and June 1994 when the number of resettlers living in the city increased from 12 000 to 25 000.[71] Their presence increased the pressure on the Minsk City Council, which was already dealing with an unanticipated growth of the city that has far outdistanced its neighbors in the modern period in terms of size. As noted in Chapter 1, one in four urban residents in the republic currently live in the capital city.

Accounts suggest that the Minsk resettlers remained an isolated and very anxious sub-community. One source observed that resettlers from Naroulya, Chachersk and Brahin regions were living in a multistorey building nicknamed 'Shanghai.' The village mentality persisted in the urban environment. Old 'babushkas' sat all day on benches at the entrance. At weekends, the volleyball pitch was transformed into a dance floor. The impression of a village was heightened by the garden plot of potatoes, beet and tomatoes that adjoined the building. The source noted that 'there is a lot of drinking here' but that such practices were understandable, and that on weekends, many of the inhabitants returned to visit their old homes in the contaminated zone.[72]

Further evidence of the plight of evacuees living in Minsk was provided in another article that began by asking where in the capital could the Belarusian language be heard most often? Not at the philology department of the university, the author replied, or the Kupala Theater, or even Independence Square during meetings of the Belarusian Popular Front, but in the micro-districts of the capital: Malinauka, Shabany, and Kuntsauyshyna. Most of those who moved to Minsk managed to get onto the list for urgent evacuation from the contaminated regions: the sick, the disabled, families with young children, the old and lonely, and those for whom 'radioactive AIDS'[73] was most dangerous. However, once they had moved to Minsk, 'our authorities considered their case closed.' The evacuees were left alone with their problems and concerns. In fact their health had not improved. One possible reason for their poor health – and one corroborated by a recent study[74] – is that many were evidently moved to the Zavadskiy (factory) district, the area of Minsk most contaminated by industrial pollution. In April 1995, the author visited the Malinauka-4 district in the southwestern extremity of Minsk. There, numerous new apartment blocks have been constructed for resettlers, who could be seen congregating, village style, in groups. The area is literally the last such construction on the city boundary. A vast plain adorned with abandoned cars is located on the other side of the apartment complex, and a giant thermal power station is the main feature of the landscape. Smoke poured forth from one of its chimneys, indicating that the new location for the resettlers may constitute a pollution hazard. Ironically they might have had a better chance to improve their health had they remained in the contaminated zone.[75]

LIFE IN THE CONTAMINATED ZONE

After the bulk of the evacuations had been carried out, those people remaining in the contaminated zones developed their patterns of existence. In speaking of the zone, the reference is to a contiguous area of Homel' and Mahileu oblasts in the southeast and east of the republic. Though the Pinsk region and areas close to Minsk are included in the fallout zone, they are considered less hazardous. The zone itself has become the subject of great public, national and international attention, and is frequented regularly by teams of western and Japanese scientists and doctors, and by members of western charitable organizations. Some Minsk residents refuse even to travel through the contaminated areas of Homel' and Mahileu, though there has been a stream of Belarusians into the area (see below). Even so, not all areas have received adequate attention. At a conference in Minsk in 1992, one of the biggest 'celebrities' was a woman who worked on a farm in the Brahin region which, at that time, had been ignored by both state and nongovernmental organizations in terms of aid (reportedly by 1994, the Japanese had begun to take an interest in its plight).

Here, briefly, the intention is to provide an overview of daily life in the zone, before turning to specific aid programs offered by Belarusian and international organizations. Hyperbole concerning the fate of the zone certainly exists. This is less a population living under a death sentence than a group that has become the focus of speculation and scientific inquiry. As noted, the main danger to this population today lies in the ground rather than the air, and the impact of radiation transferred from the soil to the roots of plants is far from definitively studied. The effect of cesium upon the human organism is the subject of a scientific study at the Institute of Radiation Medicine at the present time.[76] That the population which remained in the contaminated zone consumed and disseminated products well over the acceptable limits is not in doubt. One source notes that after the initial period of concern about the effects of Chernobyl, there followed a period of calm, and the problem faded from public view. A mistaken view prevailed that the worst was already over. In 1986–7 therefore, on the contaminated territories of Mahileu and Homel' oblasts farmers produced the following crops with above-permissible norms of radiation: 652 000 tons of grain; 116 400 tons of potatoes; 551 600 tons of hay; 828 800 tons of milk, and 28 000 tons of meat. Radiometric control over the 'cleanliness' of the food produced remained in its infancy: there were few individual Geiger counters to be found, and very rarely was local agricultural technology equipped with hermetic cabins and conditioners.[77] The situation was addressed in succeeding years, though some of the measures taken can be considered controversial, particularly the application of fertilizer to the soil on a wide scale.[78]

Though the authorities issued several decrees dividing lands into dangerous and cultivable, it was evident that many contaminated territories were included in zones designated as free from high-level radiation. One individual example, which could hardly have been isolated, was that of the collective farm called Prahres in Kastsyukovichy Raion (Mahileu Oblast). On the orders of its chairman, A. Ahayeu, 2157 hectares of land with radiation levels exceeding 40 curies/km^2 were cultivated, and the chairman ignored orders to take the land out of productive circulation. The State Chernobyl Committee attempted – the source does not reveal whether it was successful – to get Ahayeu convicted of criminal activity.[79] Up to the end of 1992, there were no regulations and no authority to prevent passage into and out of the contaminated zone, and the land within could be used without regulations. Some Belarusian residents were even attracted to such a zone:

All these years [i.e., from the spring of 1986 to the end of 1992] the zone attracted people of diverse types like a magnet, particularly the homeless vagabonds who did not fit into society but found food, water, and shelter there for 'civilized poachers.' Wild berries and mushrooms were taken out of there, as well as entire houses and buildings, construction materials, and items that belonged to the people evacuated in 1986 or who have been transferred to the noncontaminated areas. Such practices facilitated the spread of radionuclides across republican territory.[80]

There was also a constant stream of visitors to the zone in the years after Chernobyl. In 1992, for example according to the Belarusian Ministry of Internal Affairs, the numbers averaged 2400 per month, and a significant minority arrived without authorization.[81]

The collection of mushrooms and berries, which is virtually a national tradition, soon became an especially hazardous enterprise. In the summer of 1993, many families still continued this practice. Because of high retention levels of cesium in mushrooms and berries, they were not permitted to be collected in areas in which the soil's cesium content exceeded 2 curies/km^2.[82] In the Krasnapolle region (Mahileu Oblast), contamination levels remained very high: mushrooms, blackberries, blueberries and raspberries were found to be heavily contaminated with radioactive elements. Radiation levels exceeded the norm by 5–10 times and thus their consumption in any forest in the region was declared impermissible. Of some 40 sanitary and epidemiological tests conducted on mushrooms and berries for radiation content, in not a single case were the levels acceptable. In the Krasnapolle Regional Veterinary Laboratory, a similar result was produced from 18 tests. Further, a Japanese Geiger counter was used at the Krasnapolle junior high school and demonstrated that, according to comprehensive tests, all forest products contained levels of radiation 5–10 times higher than any permissible international and republican levels.[83]

In the summer of 1993, a period when radiation levels in mushrooms and berries outside the immediate 'zone of alienation' around the Chernobyl reactor were reportedly higher than at the time of the accident, warnings were issued regularly to the republican population about buying such products at markets. For example, in the Bykhau Raion (Mahileu Oblast), residents were advised not to buy foodstuffs on the street if the salesperson had no certificate indicating that the product had been tested for radiation content. They were also advised to be in no hurry to immediately consume any berries or mushrooms collected. First, it was important to check whether the latter contained radioactive substances above permissible norms at the laboratories open on the collective farm market, the regional sanitary-epidemiological station, the veterinary clinic, or at the local radiometric control centers that existed on almost all farms.[84] The warning implied that the control system was comprehensive, but in practice it is highly unlikely that this was the case.

Life in the zone remains difficult and stressful. In the spring of 1993, according to the State Chernobyl Committee, about 300 000 people were living in an area that would result in individual radiation loads of 1–5 millisieverts per annum. It was recommended once again that each of these settlements should possess a passport that indicates the socioeconomic and radiation characteristics of the towns, the medical and demographic conditions, types of occupation, levels of territorial contamination, radiological and ecological evaluations of agricultural work, and the life dosage prognosis for those residing there. Of chief concern were 200 settlements in which levels exceeded 3 mSv per annum and the regional centers of Vetka, Karma, Chachersk, Naroulya and Brahin (Homel' Oblast), all of which required urgent attention. Those settlements in which the dose load annually would not exceed 1 mSv would be allowed to continue existing methods of radiological control.[85]

In the contaminated regions in 1994, it was reported that 46 205 schoolchildren remained and that not only were their living conditions very difficult but also their education levels had begun to suffer. In the school years 1990–2, each school had a deficiency of 5–10 teachers, while existing staff taught up to double their normal weekly load. Because of a shortage of funds, schools were suffering materially too. Canteens had not been properly completed, and buildings had problems with poor water supply and inadequate sewage. Fifteen schools in the contaminated zone lacked a canteen.[86] At the other end of the chronological scale, elderly people who refused to leave their homes were penalized. Many were fearful of the possible consequences of their moving. Certainly 12 seniors moved from the village of Vepryntsy in Cherikau region died in the period October 1991 to April 1993[87] – but since there is no indication of their health status beforehand or indeed their age, one can only speculate whether the stress caused by their evacuation played any part in their quick demise.

In Okhara, by the spring of 1993, local residents who refused to leave for noncontaminated zones lost their so-called 'death benefits.' One source complained that it was a callous act to take away pittances from 'old babushkas and didas,' especially when so much money expended on the Chernobyl aftermath had been wasted, poured into hopeless schemes or into the pockets of corrupt elements. Sometimes settlements were found on the borderline between admissible and nonadmissable levels of radiation in the soil. The village of Okhara, for example, was suddenly discovered to be 3.3 curies/km^2 over the radiation limit and residents were ordered to pack their belongings and leave for a clean zone. Many refused, including the local collective farm called Ilich, whose members continued to plant wheat and fodder. Such were the divisions caused in the village that one couple went through separation and divorce: the woman gave in to the pleas of her relatives and left the village while her husband remained. This was termed a Chernobyl-style divorce by the locals![88]

The village of Volchas, in Krychau Raion (Mahileu Oblast), was the subject of an apparent official error. It contained a population of only 117, who continued to work and raise children despite being affected by increased radiation content in the soil. Until June 1992, Volchas residents received monthly bonuses for living in a contaminated zone. The bonus reportedly enabled inhabitants with modest incomes to purchase clean food products. However, the bureaucrats who prepared the draft resolution on the areas to be included for bonuses 'forgot' to include Volchas on their list. The village was thus excluded from further payments. Six months later, the State Chernobyl Committee confirmed that Volchas had a cesium content in its soil of 7.45 cu/km^2 and was thus eligible to receive a Chernobyl bonus and ordered payments to begin from the date of publication. Volchas residents were evidently incensed that they were to receive no compensation for the six months of nonpayment of their allowance.[89]

In the Brahin and Khoiniki raions of Homel' Oblast, the situation in the fall of 1993 was declared to be catastrophic. These 'remote regions' were 'heading for the socio-economic abyss with increasing speed,' according to the chairman of the Khoiniki Raion Executive Committee, Aleh Akusha. Though the whole area had been declared a possible future radiation-safe belt, no practical steps had been taken to secure such a status. Even the question of supplying gas to the regional center of Khoiniki had been left unresolved. Promised housing had not been constructed. Many specialists, including doctors and teachers, had left the area. According to Akusha, there was little doubt that the necessary positions would have been filled had the town been able to provide apartments for the medical and educational workers. Even 100 local teachers did not have their own apartments and the waiting list for housing dated back to 1975. Fuel problems were endemic. In the 1992–3 winter, part of the town lacked heating as no fuel had been supplied. Schools, kindergartens and the local hospital had been without heat. Suffering residents heated bricks on kitchen stoves and put them in corners of different rooms to keep children warm. All appeals to higher authorities had been met with indifference. Only the innate patience of the residents had prevented mass riots, Akusha added.[90]

The other regional center of Brahin was reportedly in an even worse condition. The town center, once a flourishing center of Poless'ye, had been declared evacuated. Now Brahin was virtually a front-line settlement, half-abandoned and without authorities in the region. The new group of leaders had 'run away to the mainland.' Those who had been loudest in their criticisms of the party nomenklatura in the first post-Chernobyl years were today silently leaving their executive committee offices. The Brahin residents were vocal in complaints against their local leader, a man called Prakopau, who had appeared to be a dedicated person until elected to the Supreme Soviet, after which he had visited Brahin only rarely. He had become in effect a 'paper boss,' but still received a full salary for his position in the town.[91] Towns and villages in the contaminated zone, clearly, were suffering from official neglect in the period 1993–4. A sense of hopelessness had become pervasive in many areas and there appeared to be no relief in sight.

Brahin region has also been a notable location for *samosely*, the term used to denote the predominantly elderly people who have returned without permission to their abandoned homes. According to one account, the returnees usually came back quietly, trying to avoid the local authorities and militia. A major center for returnees was the evacuated village of Sabaly, a 'dead zone' that became alive again as a result. The author of the account, a Brahin-based journalist, visited the settlement, noting that its appearance resembled the after-effects of a violent storm, with roofless houses, and crockery scattered around. He was informed by a local resident that the inhabitants lived like immigrants, unregistered, with no rights. The man and his family had been evacuated but in the new location the children had been victimized at their assigned school, and his wife, who had been employed at a livestock farm, had returned home in tears when her co-workers accused her of being 'infectious.' Winters in Sabaly were cold. The wooden hut in which the family lived had no heating system and the children had fallen sick. The man's wife suffered from nose bleeds and high blood pressure: 'radiation levels are probably very high here.' Nevertheless food supplies were plentiful, with fat, milk, potatoes and eggs. Fishing also proved profitable, and even the pigs were fed on a diet of fish. The main dilemma of the *samosely* in this village was less their illegal presence than their inability to receive any form of aid. The closest state farm was reluctant to assist them, the manager remarking that they had chosen to return of their own free will. Some *samosely* remained permanently in the evacuated village, while others treated their former homes like a dacha, arriving for periodic visits.[92]

In the spring of 1995 it was possible for the author to undertake a tour of the contaminated regions in the area around the town of Chavusy, Mahileu Oblast, some four hours by bus east of the city of Minsk. The level of cesium contamination in the soil varied from 1 to 15 curies, though local observers claimed that most areas had received from 1 to 5 curies. Hence in contrast to regions such as Brahin, the area in question was one of periodic control and of lower-level radiation. As such it might be described as a forgotten zone. A local flax combine had been closed down, along with several enterprises based on dairy products. Many of the wells on the village streets had also been boarded up. In the local store in Chavusy itself, food supplies were minimal (especially when compared with the city of Mahileu) and prices high. The consumers could choose items from brown bread and dry sausage to cucumbers, pickles and an unidentifiable form of fish. There was no cheese, little butter and not even vodka was available.

I visited five homes in the suburbs of Chavusy, and in the nearby villages of Kam'yanka and Vileyka.[93] Clearly the lives of the residents had been adversely affected by Chernobyl. Since compensation payments for living in the zone had been eroded through inflation, the ability to obtain clean food supplies from outside the region had been greatly restricted. Most of the families were dependent upon personal livestock for survival (though technically each belonged to a single collective farm, the activities of the latter appeared to be limited). The overwhelming impression was one of acute poverty to a degree not seen in the noncontaminated regions of the republic. Such poverty was not universal – one single mother with a 16-year-old daughter appeared to be relatively prosperous – but it was the predominant feature.

A second characteristic among these families was alcoholism among one or both parents. Evidently many families were typical in this regard. Often the father was the alcoholic, though at one farm both parents were declared to suffer from this syndrome. The farm in question, in Vileyka village, offered a scene as pathetic as any conceived by Dickens, with a family of seven children living in a single room in a wooden hut, almost suffocated by the fumes from the (broken) log fireplace. The grandmother lay in one of the four beds in the room, having broken her leg while out in the garden. A four-year-old girl fed herself with spoonfuls of dried baby-powder, while the 11-year-old boy had never attended a school and was in a filthy state. There was no food in the hut, no running water and no other amenities; and the family possessed no livestock. These families may indeed have been victims of Chernobyl, but principally in the sense that acute despair may have driven the parents to the 'demon vodka.' Alcoholism in Soviet and former Soviet villages is hardly a new phenomenon. What was surprising was its prevalence in the Chernobyl zone nine years after the event. The villages visited had no roads, no inside toilet facilities and no hot water; they resembled something out of a past century.[94] The Chavusy region may have suffered more than most other contaminated areas from lack of attention by government or charitable organizations because of its relative remoteness. Because they are better known as the more heavily affected areas, Brahin, Khoiniki and other zones of acute contamination receive far more attention from international scientific, medical, and charitable organizations.

Even though by 1992–3, the bulk of the decontamination work had been completed, living inside the zone or even just outside it still presented its hazards. In the town of Valozhyn (Minsk Oblast), for example, which officially is not in the zone of contamination (though part of Valozhyn Raion to the east and southeast of the town is within the zone), problems had been caused by the constant movement of vehicles from other cities through the center of the town, bringing radioactive dust on their tires. Apparently to stand on the sidewalk for a period of 10–15 minutes was to be convinced that the main street was a military thoroughfare during army maneuvers. Incidentally, the town was suffering from a spate of thyroid gland disorders and local residents were attributing these ailments to the spread of radiation.[95] The town 'needed to be saved.' The local Pyarshai Hospital, however, was suffering formidable problems: old walls, a leaking roof, and heating and sewage systems that did not work. The complaint from Valozhyn Region was that the hospital was receiving more aid from the Germans than from the Belarusian authorities: German charities had donated a Mercedes, an ambulance, an electrocardiograph machine and three deep-freezers. The region contained 102 settlements with significant fallout from Chernobyl and, by the spring of 1993, only one – the tiny Rodz'ki with six farmsteads – had been evacuated.[96]

Another difficulty for zone residents was dust storms, which could cause the rapid dissemination of radioactive particles without warning.[97] The results of Chernobyl and the failure of the authorities to improve sufficiently the living conditions in the contaminated areas of Homel' and Mahileu oblasts had not surprisingly led to a drop in the population of these regions. It had also resulted in a significant and as yet rarely analyzed role of international charitable organizations in the contaminated zones. Though received with something less than enthusiasm by the Belarusian authorities,[98] Belarusian officials recognize – as has been demonstrated here – that left to its own devices the state simply cannot deal with the repercussions of Chernobyl.

AID TO BELARUS: THE ROLE OF INTERNATIONAL CHARITABLE ORGANIZATIONS

Aid to Chernobyl victims has become a controversial and much debated issue. Among the questions most frequently asked are the following: Is the aid useful? Does all the material aid and medical equipment reach the intended recipients? Why has there been so little coordination between the various humanitarian groups? A reporter from Belinform added other questions: why is it necessary to have 38 such organizations within one republic? Why do children continue to die despite such efforts? Why are there endless lines at the medical institutions dealing with Chernobyl? Why have some children visited foreign countries several times while others have never been given such an opportunity?[99] These questions and others have supplemented actions by the government of Belarus, which has tried to hinder the work of such organizations at every opportunity, partly because the majority of them were initiated and continue to be operated by members of ostensible 'opposition' groups.

There have been a variety of criticisms of humanitarian aid and the alleged abuses that can be perpetuated by such assistance. A few examples will suffice. Some citizens manage to get classified illegally as resettlers, one source noted, by visiting the Brahin region for about 20 minutes: sufficient time to mark the papers for a business trip, after which they become 'liquidators' and enjoy additional holidays and visits to Spain and the Netherlands for treatment.[100] A vicious assault on humanitarian aid from another source maintained that it has created a consumers' attitude toward society. Psychologists had diagnosed this as a situation in which people like to consider themselves victims, and have no shame to indulge in begging. The source made it plain that the allusion was not to real victims of Chernobyl, but to those who live all their lives at the expense of the state, and refuse to take any initiatives themselves.[101] Clearly such people exist in Belarus, but they were hardly created by Chernobyl or humanitarian aid; rather their origins may lie in the years of Soviet rule. The humanitarian groups, for the most part, have undertaken important and selfless work for the benefit of Chernobyl victims and for Belarusian children generally.

It is important, nevertheless, to make some distinctions between the various aid groups. Some began purely as Communist fronts, often hoping to receive significant hard currency donations from the West. The name 'Children of Chernobyl' was so widely used that it was often difficult for humanitarian groups from the West to determine which organizations comprised grassroots leaders and which were established by the government. The vast majority were formed in 1989–90, a period when the widespread contamination from Chernobyl was well known, but a time when the overall coordination of relief efforts was very much centralized in Moscow. Thus it was a difficult task to obtain material support from the West since this generally was directed through the Soviet capital.

Though many organizations developed in Belarus, very few could be described as financially viable. Most in any case could operate only with links to Western charitable associations – such as 'Citihope International' – which provided hard-currency funding for operations. In the spring of 1992, a sociological research association conducted a survey of 826 respondents living in the following contaminated regions: Brahin, Chachersk, Buda-Kashaleuski, Mazyr (Homel' Oblast); Slauharad, Cherykau, Krasnapolle (Mahileu Oblast); Luninets Raion (Brest Oblast); and the village 'Druzhba' in Pukhavitski Raion (Minsk Oblast), a settlement constructed for evacuees from the contaminated zone. When asked which organizations had been the most effective in providing aid to affected regions, a substantial 64 percent cited the Belarusian Charitable Fund 'For the Children of Chernobyl,' 11 percent cited government organs; 10.5 percent raion councils and executive committees; and 9.3 percent committees of the Supreme Soviet. Other charitable funds, such as the Fund of International Cooperation for Social Protection 'Belarus' and the Masherov Belarusian Union 'Znich' received smaller percentages of the vote.[102] One group has thus dominated the public's perception of domestic humanitarian associations. Why is this the case?

The Belarusian Charitable Fund 'For the Children of Chernobyl' (BCFCC) was created in November 1989 as a nongovernment committee initiated by members of the Belarusian Popular Front (BPF), led by Gennadiy Grushevoy, a parliamentary deputy with a doctorate in philosophy and logic. In the fall of 1989, a so-called 'Chernobyl Road' had taken place as a protest action. Grushevoy and another organizer, Yuriy Khodyko, were arrested, though they were detained only briefly by the authorities. By the end of this year, the committee had sent its first group of children abroad. Selected from Khoiniki Raion, Homel' Oblast, they were sent to India. By early 1990, agreements had been signed with the governments and local authorities of Poland and Norway to send similar groups to these countries. Though his Committee was still not recognized by his own government, Grushevoy had soon secured agreements with six countries, and some 5200 children were sent for periods of recuperation abroad. By the fall of 1990, a program of aid to sick children had been established and contacts made with a variety of public and religious organizations around the world. Finally, in November 1990, the legal status of the committee was formally recognized as the Belarusian Charitable Fund 'For the Children of Chernobyl.'[103]

The Fund soon expanded to encompass a membership of some 5000 volunteers, each of whom had specific responsibilities. Initially, its chief task was to send children for periods of recuperation abroad (see below). In 1994, prior to the election of Lukashenka as president, it had 56 grass-roots organizations (3000 workers) distributed throughout the republic, and contacts and partnerships with more than 20 countries, but somewhat dominated by Germany.[104] The Fund from the outset was regarded with suspicion and even hostility by the health authorities of the republic, though it had a good working relationship with the Minsk City Council, and the areas of health that fell under city rather than republican jurisdiction. Generally, however, in the period 1990–94 it operated in an atmosphere of dissidence and isolation that was heightened by the relatively weak position of democratic political forces in the republic.

The Fund's leader, Grushevoy, regards Chernobyl as one of the agents not only of political change in the republic, but also as an event that forced the public to reassess its attitude toward the state. In his view, Chernobyl demonstrated to the people that the government was incapable of improving the situation so that contaminated areas became safe for human habitation. The Belarusian authorities hoped to receive the necessary orders and funds from Moscow, he maintains, and would then draw up a program based on regulations set forth by Moscow. Today,[105] however, it was necessary to change the principles and structure of activities. What should be the role of a new government? In Grushevoy's opinion, the government must support the initiative of the grassroots community; it should bring about conditions in which people are encouraged to develop their own initiatives. The government of Belarus had not adopted such a role. Rather it perceived a threat to the traditional image or way of thinking that people have about such matters. Ministries were reluctant to witness the elimination of their authority. Yet when people attempt to resolve their own problems, and receive funds for their endeavors, then it becomes impossible to control them with orders. Only through the legal system can their activities be subordinated to those of the state. 'Where the law governs, the power of the individual ends.'[106]

Grushevoy possesses a singular view of how children in the republic should be assisted, that is not based strictly on either the radiation or the health situation. In his view, the Chernobyl disaster was a unique psychological phenomenon that affected all children of the republic, even those who have not suffered from fallout. In sending children abroad for periods of relaxation, therefore, one need not discriminate between those who are sick and those who are healthy. All are in equal need of recuperation. By 1992, the Fund had listed in a computer bank 17 000 children who had traveled abroad for recuperation. It relied mainly on the work of public organizations in the localities for making its selections. This number represented approximately half those sent abroad by all charitable organizations at this time. In 17 regions of the Chernobyl zone, the Fund had also established cooperation with foreign partners for the operation of hospitals, schools and kindergartens.[107]

The Fund's activities have expanded since 1990 to cover the medical care of children within Belarus. It has provided medical equipment to 26 district hospitals and has been responsible for sending 200 children with serious health problems to different hospitals. By the spring of 1994, the Fund was involved in the resettlement program, the health effects of the disaster, the rehabilitation of children, and education of professional people involved in the aftermath of Chernobyl, such as doctors, farmers and construction workers. Aside from Germany, its other main partners were Austria, Italy, Britain, the United States and Canada.[108]

For over two years, the Fund concentrated its efforts on another ostensibly Chernobyl-related health problem (discussed in more detail in Chapter 4): children's diabetes. It provided insulin as well as education on the subject – often through brief sojourns in hospitals abroad – for Belarusian doctors. One such venture was a project with British specialists, with financial support from the British Council, which began in October 1993. Two educational programs had been created: the first was in Homel' Oblast for a speciality clinic within the framework of the Homel' Health Department. This program was under the jurisdiction of the Homel' city government, the British specialists, and the Fund. The second was located in the city of Minsk, at the Institute for Medical Refresher Courses, where medical management courses were scheduled to begin in 1994. Grushevoy commented that such programs had three basic purposes. First, they enabled republican doctors to gain practical experience to complement their theoretical knowledge of medical problems. Second, they established contacts between Belarusian and foreign doctors that constituted 'an unbreakable chain of information' and could be furthered through conferences and meetings. Third, and less concretely, they modified the nature of the current health system, which hitherto had suffered from an excessive dependence on the state. In fact, according to Grushevoy, the state had begun to recognize the benefits of such programs, and the Fund had tried to recruit state specialists into such schemes in order to broaden their education and their general approach to medical questions. In some cases, cooperation was short-lived. The chief endocrinological center of Minsk, for example, Children's Hospital No. 3, was equipped with foreign technology by the Fund for a period of some months, after which cooperation was ended abruptly in 1992, on the grounds that its staff were not sharing their benefits and information with other centers.[109]

The concept of humanitarian aid with regard to the aftermath of Chernobyl is a complex one. There seems to be a consensus that such aid is needed, but there are differences over the form it should take, and the degree to which the various organizations should cooperate with one another. An American who has worked extensively in Russia and Belarus for World Vision International commented in the spring of 1992 that he saw two areas that were in need of improvement. First, in his view there was a misapprehension in Belarus (he was speaking to a Minsk audience) that international Nongovernmental Organizations (NGOs) could only work with a single organization. In Africa, Latin America, and Asia, however, the goal of US NGOs was to cooperate with as many charitable organizations from the needy country as possible, because each possessed different strengths that could be harnessed. Second, he outlined what he perceived as 'negative competition' that sometimes seeks to destroy the work of other organizations that are also performing important tasks. In Belarus, the human needs were great, thus it was essential to develop partnerships.[110]

In Belarus, however, the disparity between the charitable groups in terms of financial support and material possessions was considerable. As their estimated budget is difficult to decipher in stable currencies, the following table has them divided into three categories: healthy, stable, and weak, and selects only the top 15 organizations prior to the Lukashenka presidency.

In the summer of 1994, two other nongovernment funds were cited as providing significant aid for programs on which children travel abroad for recuperation in addition to the above list: 'Independent Aid for Children' and 'For Life After Chernobyl.'[111] In general, nonetheless, the number of financially healthy humanitarian organizations prior to the presidency of Lukashenka was small.

The aid has taken various forms. One of the most significant projects has been set up by the Swedish firm Smelt-intag, to introduce a Center for Radiation Medicine in Homel', with 475 beds and a polyclinic that will permit treatment for 500 people on a periodic basis. The project was scheduled to be completed by the summer of 1995, replete with the most up-to-date equipment for the diagnosis and treatment of those suffering from the Chernobyl disaster.[112] Another scientific program has emanated from the US firm Lata, in cooperation with Mitsui of Japan. The goal is to

Table 3.4 *Major charitable organizations in Belarus, 1989–93*

Rank/Name	Status
1. Belarusian Charitable Fund 'For the Children of Chernobyl'	Healthy
2. The Fund of International Cooperation for Social Protection 'Belarus'	Healthy
3. 'Protection from the Atom' Benevolent Fund	Stable
4. Belarusian Committee 'Children of Chernobyl'	Stable
5. International Humanitarian NGO 'Chernobyl Aid to Belarus'	Stable
6. The Belarusian Social-Ecological Union 'Chernobyl'	Stable
7. Masherov Union 'Znich'	Stable
8. Republican Association of Citizens' Legal Protection Organs	Stable
9. Belarusian Operational Committee 'Echo of Chernobyl'	Weak
10. Belarusian Republican Charitable Fund 'Life After Chernobyl'	Weak
11. 'Chernobyl' Union	Weak
12. Belarusian Union of Participants in the Liquidation of the Consequences of the Chernobyl Catastrophe 'Pripyat'	Weak
13. Belarusian-Japanese Society 'Chernobyl-Hiroshima'	Weak
14. Belarusian Fund for the Invalids of Chernobyl	Weak
15. Belarusian-German Cooperative Charitable Enterprise '21st Century Hope'	Weak

Source: Narodnaya hazeta, 20 July 1991; Chekha, 1993.

study the contaminated territories of Russia, Belarus, and Ukraine and examine the migration of radionuclides from the soil into the food chain; to assist with the control over the dissemination of contamination-free food products; and to assist with medical treatment and sanitation conditions of the people living in the contaminated regions. The Japanese firm is reportedly willing to contribute up to $80 million on these projects.[113]

Evidence of international assistance for medical and scientific projects is visibly apparent in Belarus, with foreign equipment in many of the major hospitals, truckloads of goods from Germany outside the German Embassy and in centers such as Minsk and Homel', and a semi-permanent presence of charitable organizations, such as Citihope International, Children of Light, World Vision International, and Friends to Friends.

Homel' has inevitably been the focus of most efforts. In the period between April 1986 and September 1993, 125 health care facilities were established in the oblast using funds assigned officially for Chernobyl repercussions. These facilities were able to purchase 120 ultrasonic devices and 32 'super-high frequency' devices. In addition, a radiation medicine prophylactic center, a medical college and a Belarusian-Dutch polyclinic had also been opened. Two diagnostic centers – the Sarakava Foundation and the Red Cross International League – had been created, and the World Health Organization had developed a program for the Homel' Medical Institute. Finally, in Homel' Oblast, an agreement was signed to set up a Holland-Homel' Project; and this contemporary recipient of high radiation fallout was said to be receiving considerable attention from Japanese specialists from Hiroshima and Nagasaki.[114] The danger in Homel' might be the plethora of aid programs and new medical institutes. The contaminated area has become a vast human scientific experiment that has aroused interest among scientists worldwide. It is fortunate that most of these experiments and aid programs do not deal directly with the official authorities. On the other hand, they do lack a centralized coordinating body. The question that arises is whether the various groups will publicize their findings so that they are able to work in conjunction with rather than parallel to one another.

A typical example of a local organization's work was offered in the spring of 1993 by Georgiy Vecher, the vice-president of the International Humanitarian NGO, 'Chernobyl Aid to Belarus.' According to Vecher, the strongest links with international organizations were those with southern Germany, particularly the cities of Heidelberg and Würzburg. Children from Belarus travelled there for a period of relaxation, while the Germans contributed medical equipment and food products to Belarus. Relations were also declared to be 'good' with Italy and Britain. In March 1993, 13 children from Brest Oblast had travelled to Britain for treatment. Since such travel was expensive, a number of Belarus firms had offered financial support, including a dairy products factory in Salihorsk, a Minsk metallurgical works, and a tape factory in Mahileu. The Peace Fund of Belarus had also offered aid in securing from abroad over the previous year almost 13 million tons of food prod-

1. (above) Statue of Lenin,
    Independence Square,
    Minsk

ПОДВ   НАР

2. (left) Victory Square, Minsk
    (courtesy of Esther Van
    Nes)

3. (left) Parliament buildings, Minsk

4. (below) Independence Square, Minsk

5. Victory Day parade, 9 May 1994 (courtesy of Esther Van Nes)

6. Supporters of the Belarusian Popular Front in Minsk, 1994 (courtesy of Esther Van Nes)

7. Veterans of the Great Patriotic War, Minsk (courtesy of Esther Van Nes)

and 9. Communists at the May Day parade, 1994 (courtesy of Esther Van Nes)

10. Village of Kam'yanka, Mahileu Oblast

11. Residences constructed for evacuees, Malinovke–4 district, Minsk

12. Family in contaminated zone, Chavusy region, Mahileu Oblast

13. Roman Catholic Cathedral, Minsk (courtesy of Esther Van Nes)

ucts, and over 40 million tons in humanitarian aid. The Japanese had sent Geiger counters valued at 2 million yen.[115]

How many children had been sent abroad by the Peace Fund of Belarus and how were they selected? In the previous year, responded Vecher, over 2000 children had been encompassed by the program. Activists in the localities had selected the children, but there were also some who were chosen by the host families themselves. The NGO was unable to make prognoses about the future; it operated in hours rather than days because expenses were mounting, but it was planning to hold a conference on the medical aspects of the Chernobyl accident. The above operation is a small one: in contrast, the Charitable Fund 'For the Children of Chernobyl' dispatched over 60 000 children to Germany alone in the summer of 1994. Germany appears to be the major participant in such programs, though a wide variety of countries has taken part, including Poland, Slovakia, Cuba, Israel, Belgium, Britain, France, Holland, Spain, Italy, the United States and Canada.[116]

How useful are such programs? The observer, faced with this question, has few sources to guide judgment. The programs seem to be highly popular with the children themselves, though the psychological effects of returning to territories that have fewer resources and which may also be contaminated must be considerable. For the most part, the children who travel are not those who are sick; indeed an illness could preclude travel. One can thus refer to Grushevoy's claim that all children have suffered emotionally from an event of such magnitude – and the fear of radiation remains a very real factor in inducing stress – or else raise questions about the wisdom and effectiveness of such programs, other than that they can act as a remarkable educational process for those involved. Certainly there is an absence of logic involved because those children sent to countries that require transatlantic flights receive more natural radiation from the plane trip than they would had they remained at home.

Summer visits abroad for Chernobyl children are also often an embarrassment to the government because they draw attention to the inability of official organs to provide such aid (initially, the government managed to organize several such visits, mainly to Cuba and other Communist countries).[117] Grushevoy is well aware of the limitations not merely of government aid, but of the sort of mentality that pervades republican leaders:

The current government consists of people whose main drawback is their lack of professionalism. There is a more profound defect of this government. These people formed their experience of state government under the most extreme administrative-command system. The overall potential of the government is limited by the previous experience of the perestroika and pre-perestroika periods. In reality they cannot act any other way. They have tried to adapt their old ways of ruling the country to the requirements of the present, and as a result there is a deformation of real processes that are occurring today ... They are tied to their habits. They apply old methods to new conditions and become even worse than before.[118]

The government, according to this outspoken critic, is not capable of reacting to the dilemmas caused by radiation contamination. We have demonstrated above the shortcomings of official aid. In addition to initiative, the government is chronically lacking in funds. An example of how even the state aid programs are dependent upon hard currency from abroad was provided by Georgiy Lepin of the Belarusian Union of Chernobyl, affiliated with the Ministry of Internal Affairs. In September 1992, he commented on the proposed construction of sanatoria for children suffering from Chernobyl. One such sanatorium was said to cost 25 million German deutschmarks, and the Union was declared to be reliant on German aid for its construction, since that figure was well beyond its means.[119] Materially, therefore, the Belarusian state has never been in a position to address the enormous effects of a major accident.

Further, the nature of the present state, its demographic and social history, preclude an objective approach to such a problem. Observers in Minsk draw a contrast between the relative efficiency of medical institutions administered directly by the city, and those under the control of the republic. The latter are considered to be poorly run and generally unsatisfactory. The singular lack of reform in Belarus – indeed the significant opposition to change and even support for the restoration of the Soviet Union – reflect the lack of maturity of the present state and the continuing iron grip upon state structures of the former Communist Party nomenklatura. As Grushevoy points out (see above), such people have not changed their pattern of thinking as a result of the independence of Belarus.

Because much of the old elite is still in authority, and has even been somewhat rejuvenated by younger figures, there is as yet no possibility of a social and economic catastrophe being dealt with in national terms. The response to Chernobyl at the official level represents a series of half-steps, a reluctance to commit resources to the full, and a continuing resentment of the foreign 'intrusions' into the republic to provide medical, material, and even spiritual aid. The state's attitude has been reinforced from another quarter: the viewpoint expressed by some of its own medical leaders and supported by influential international experts that the effects of the accident in terms of health repercussions have been exaggerated. If there is in reality no significant health crisis in the republic, then what is the purpose of the plethora of charitable groups; visits of children abroad; increased medical attention; and the like? The propagation of this line – its authenticity will be examined in the next chapter – permits the state to avoid commitment to Chernobyl and concomitantly to question the need for the existence of charitable groups, particularly those that have been established by vocal critics of the government and the president. Let us therefore turn to the question of the health consequences of Chernobyl in Belarus, as manifested in the first post-Chernobyl decade.

4 Medical Consequences of a Nuclear Disaster

INTRODUCTION

The question of the medical consequences of Chernobyl has elicited more discussion and resulted in more controversy than perhaps any other single issue. Concern for health has been the reason for the constant visits of scientists and medical personnel to the contaminated regions; to the hospitals and clinics of Kyiv, Moscow and Minsk; for the involvement of organizations such as the International Red Cross, the World Health Organization, Greenpeace International and other groups. It was the reason behind the International Chernobyl Project, commissioned by the International Atomic Energy Agency in 1990–1, which made a study of villages within the contaminated zone and published the results in the summer of 1991. Since 1986 there have been annual and biannual conferences on the topic of the health consequences of Chernobyl and the likely repercussions to be expected in the future.

Even in 1986 scientists speculated on the potential number of malignant cancers among Chernobyl victims, and on which other types of illness might develop. There have been discussions also about the number of direct casualties of Chernobyl and assault on the official and misleading figure of 31 dead (28 from radiation sickness) that was introduced and perpetuated by the Soviet authorities to demonstrate that the disaster was not as bad as initially feared and that they had succeeded in limiting its consequences. Conversely there have been uncorroborated statements about the number of dead that can be termed irresponsible, particularly among the cleanup crews. The central question that emerges after a decade is where does truth lie? How many people actually died? What are the principal results of the radiation fallout among the population today? Has there been a significant rise in leukemia? In other types of cancer? What is the state of the health of affected children? Is the Republic of Belarus indeed facing a health catastrophe, a health crisis, or merely some additional health problems in the mid-1990s?

Our comments are necessarily confined to Belarus which, as shown in Chapter 3, received the largest portion of the radiation fallout, though it should be borne in mind that health consequences were also significant in

Ukraine and Russia. After the dissolution of the USSR (and even by 1989) the Belarusians had developed a wide network of organizations that were studying the effects of Chernobyl. On the scientific level, the principal body was the Institute of Radiation Medicine affiliated with the Ministry of Health. It has 15 institutions within the health ministry, in addition to four in the Ministry of Education, 11 in the Academy of Sciences, and two in other departments.[1] The Ministry of Health itself, under V. Kazakov until January 1995, has been very much involved. The City of Minsk possesses its own medical network under a Chief Physician, which has conducted some significant surveys of population living in polluted territories within the city.[2] Homel' region also has its own medical network and is the center of a series of experiments, some of which are administered from Minsk, some locally, and some by international organizations.

It should be said at the outset that medical data are not always easy to obtain. As in any scientific inquiry, researchers are often anxious to complete their studies before releasing any results publicly. There are inevitably many disputes, sometimes developing into outright conflicts. Quite often, the same or similar statistics are revealed by quite different sources that have evidently never collaborated or consulted with each other. Sometimes the head of a center is quite willing to divulge information to foreigners and is prepared to meet on more than one occasion, whereas another might constantly avoid an interview or continually postpone meetings. Gathering accurate information is therefore a difficult task. Indeed the quest at times resembles that of a detective uncovering a mystery murder, though in this case it is often the cause of the murder that is lacking.

What has made the health situation difficult to discern in the Belarusian case is the relatively backward state of health facilities and the fact that the country has experienced an economic crisis that has greatly affected the basic standard of living. In addition to problems in health care, there arises the question of nutrition. If the basic diet is a poor one and the population generally has difficulty in purchasing nutritious food because it is well beyond the average budget (though such food is available in the major population centers), then the health situation, regardless of the effects of Chernobyl, can be expected to deteriorate. Financial compensation for Chernobyl victims had become meaningless by late 1994 because the sum allocated monthly for living in the contaminated zone had been overtaken by inflation.[3] Thus economic hardship cannot be eliminated as a factor in increasing health problems; on the contrary, it has rendered these problems much more complex.

GENERAL OVERVIEW

The prognosis for the territories affected by the Chernobyl disaster by the former USSR Ministry of Health was serious but hardly catastrophic. It was estimated in November 1986 that in the period 1986–91 one could anticipate about 1300 cases of development of hypothyroidism and sicknesses of the thyroid gland, particularly among children in the Homel' Oblast of Belarus, and the Kyiv, Zhytomyr and Chernihiv oblasts of Ukraine. Over a 70-year period, the ministry forecast an additional 600 cases of leukemia and approximately 4000 additional fatal cancers.[4] One could expect an increase of oncological illnesses for a 10–15 year period, while over the 70-year period one could expect the additional births of some 3000 children with congenital defects in development. Overall the 75 million people living in the European part of the USSR with an increased radiation background would suffer 40 000 additional fatalities in the period 1986–2056, including 5000 leukemia. Of the 40 000, external irradiation would result in 5000 deaths; internal irradiation of the thyroid gland, 2000, internal irradiation of the whole body by cesium, 33 000 cases. Additional congenital abnormalities could compose as many as 23 000 cases, or a rise of 0.3 percent. Additional oncological sicknesses that were fatal would likely rise by 0.4 percent, or eight cases per one million people per annum.[5] This prognosis was based on inaccurate statistics regarding radiation fallout, since such figures were not available until 3–4 years after the accident. Nevertheless they may serve as a guideline to what was anticipated in some scientific circles.

The first detailed outline of the problems being faced in Belarus to an international audience was a remarkable statement by the then Minister of Foreign Affairs, Pyotr K. Krauchanka at the 45th Session of the United Nations General Assembly in New York on 23 October 1990. Krauchanka was responding to an IAEA report on nuclear power operations. While hardly the most scholarly account of the situation in Belarus, Krauchanka's report was probably the most significant in terms of explaining the feelings of the public toward the radiation situation. The population, he declared, was in a 'complex psychological state,' and had begun to lack trust in official structures – particularly those in existence prior to 1986. Only after four and a half years was it possible to penetrate the 'wall of indifference, silence, lack of understanding.' Every family in Belarus and Ukraine was obliged to construct their lives around the maps that designated the known areas of high radiation.[6]

Already, Krauchanka stated, the health consequences of Chernobyl in the republic were becoming apparent. A particular dilemma was posed by children's thyroid cancer, rates of which had doubled in the southern regions of Belarus. In the contaminated zones, the number of anemia cases had risen by 7–8 times, chronic pathologies of the nasopharynx[7] by 10 times; and defects in development of newborns by 1.5–2.0 times. Changes in the various systems of the human anatomy – the immune system, endocrine, blood-forming and nervous systems – had resulted in a 'radiation AIDS,' i.e., a lack of resistance to all forms of disease. Cancer and leukemia generally, he added, had risen among children, though the peak in cancer cases could be expected in 1994–6. He was afraid that a threat existed to the 'genetic hereditary identity' of the nation to the extent that in the future, as a result of the impact of radiation, Belarusian society would see a group of outcasts in marital and other human relations. Such sentiments were already being reflected in the negative birth rate of the Belarusian population.[8]

Krauchanka's dramatic depiction of the devastation caused by the nuclear accident to his country constituted a turning point in the official reaction to Chernobyl. It was also an admission that state resources alone would be inadequate to deal with the health consequences over the ensuing years and decades. Using the pretext of the role of the IAEA in nuclear disarmament and nonproliferation, he applied it in the wider context of international aid. The world community, in his view,

> cannot step into the 21st century with clear conscience, having not solved global problems and particularly those related to the prevention of wars, the elimination of famine, diseases, underdevelopment, problems of saving people affected by Chernobyl – Russians, Ukrainians, Byelorussians, other nationals – the problem of eliminating the threat to the hereditary identity of the nation.[9]

The survival of the nation, in Krauchanka's view, was at stake. Whether such a statement represented diplomatic hyperbole or reality is a moot point. What is significant is that it emanated not from a frightened peasant community of Homel' region, but from an experienced international statesperson speaking in a prestigious setting. By 1990, the republic, though still part of the USSR, had taken a very definite step away from a Union perspective of Chernobyl and had begun to focus on its own problems.

In Belarus, by 1991, there were already reports of health problems that appeared to be related directly or indirectly to Chernobyl. One source observed in this year that the principal problem in the contaminated regions at that time was psychological stress. Studies of the all-Union Center for Psychiatry had taken place since 1986, examining 25 000 individuals in areas with different levels of contamination. A remarkable 95 percent reported some psychiatric deviations or stress reactions. Residents of the affected areas were said to be suffering constant mental stress as a result of the fear of the potential impact of increased radiation on the health of family members, particularly children. The stress factor was increased by the need to introduce radiation protection measures, and by the concomitant social and economic problems.[10]

In the following year, medical specialists in Belarus, including the Belarusian Minister of Health, Kazakov, concluded that an analysis of the state of health of children in the Homel' and Mahileu regions indicated a rise in general morbidity. The most frequent diseases among children were 'neuropsychic' disturbances, various types of 'dystonia,'[11] anemia, chronic[12] respiratory system pathology, and chronic diseases of the digestive system.[13] A comparison by the Institute of Radiation Medicine in this same period of the state of health of children living in contaminated zones compared to those in 'clean' zones also revealed problems. It reported a rise of diseases relating to otolaryngology,[14] anemia, chronic gastritis[15] and other digestive diseases by 40–80 percent in the heavily irradiated as opposed to the clean areas. In 40 percent of schoolchildren in the contaminated zones, 'functional breaches of the cardiovascular system'[16] were revealed. Such information again does not necessarily indicate a crisis, but it does suggest a distinction could already be made between clean and contaminated zones of the republic.

The emphasis in medical reports of the areas affected by high levels of radiation in the period 1990–2 is on the abnormal state of affairs in the health of the population, and particularly of children. Thus in 1991, scientists examined 739 children aged three to fifteen from the Homel' (an area in which cesium content in the soil averaged 40–47 cu/km^2); Mahileu (63.2–146.5 cu/km^2); and Minsk oblasts (5–8 cu/km^2), comparing them with children from a control region. The children in the affected regions were found to be suffering from a large number of 'asthenic'[17] symptoms (20–25 percent of the total) and 'functional murmurs'[18] (36.6–42.5 percent). The percentage of normal electrocardiograms was very low (25.8–36.6 percent).[19] In another report of 1993, there was revealed a notable rise of mortality from neoplasm.[20] In 1985, the figure was reported to be 148.2 per 100 000 population; in 1986, 156.9; and in 1989, 169.4. Also medical specialists had detected a rise in hypertonic[21] sicknesses, heart disease, and anemia in children.[22]

Similar results can be found in reports from the Institute of Radiation Medicine. It was noted, for example, in a 1992 publication that there had been a 'significant increase' in the total morbidity and anemia among pregnant women. In Mahileu Oblast, anemia were found in 9.4 percent of deliveries, a rise of five times from the pre-accident period.[23] In this same report, the Institute of Radiation Medicine also stated that the birthrate of children with congenital malformations was 15–25 percent higher in contaminated areas than in clean regions, and that congenital abnormalities accounted for 30 percent of child mortalities. However, Ihar Ralevich, a deputy chairman of the State Chernobyl Committee, declared that the years 1987–9 constituted the peak of congenital malformations, and that predictions of severe hereditary problems had not thus far been confirmed. Nonetheless, he appended, it was still too early to celebrate.[24]

One of the major difficulties pertaining to a study of health effects is that the precise cause of an ailment cannot be determined definitively. There is considerable disagreement as to whether the effect of radiation has been to increase the number of illnesses generally among the population. One or two examples may suffice. In November 1993, the Chief Physician of the Children's Hospital in the town of Svetlahorsk, L.V. Sanets, declared that the number of sick children had risen over time. The number of children with congenital defects had grown. In 1986, they were found in 0.9 percent of the 1579 children born in that year, whereas in 1992, the corresponding figures were 2.4 percent and 1351. The birthrate had fallen and the congenetic problems had worsened.[25] In contrast, an international conference entitled 'Health Problems Seven Years After the Chernobyl Catastrophe,' held on 1–2 September 1993 in the city of Homel', concluded that in the post-Chernobyl years, no sicknesses had been revealed that could be directly linked to the effects of radiation, and no overall rise in morbidity had been detected.[26] In 1994, Health Minister Kazakov attempted to put the situation into perspective. For those living in the contaminated zone, the number of general pathologies had risen, but he did not link this increase exclusively with the radiation factor. Health, he observed, is a complex of many factors, and some of them – notably living conditions and general nutrition – had worsened considerably of late. Such factors had to be included in the general protective forces of the organism; the lack of which would result in the rise of pathologies, including infections.[27]

A leading specialist on Chernobyl in the republic provided a slightly but not fundamentally different perspective from that of Kazakov. Academician E.P. Kanaplya, the Director of the Institute of Radiobiology, stated in a 1993 interview that he personally considered the increase in the disease rate in the contaminated areas was caused by the radiation factor, and that the existence of other factors aggravated the situation. A rise in the rate of infectious diseases had been accompanied by diseases of chronic inflammation, which was atypical of illnesses associated with radiation. An evident reduction in effectiveness of the oncological auto-immunity system had resulted in a large increase in different diseases of inflammation.[28] In short, radiation should be seen as a contributory factor in the rise of diseases but not necessarily the determining factor.

This explanation might assist in comprehending a notable rise in cases of diabetes mellitus[29] in the republic, which doctors in Minsk Children's Hospital No. 3 attributed in August 1993 directly to the Chernobyl factor.[30] It was reported that children of an exceptionally young age were being found to have this disease, which ostensibly is not associated with radiation. Overall incidence has also risen sharply. Prior to the disaster, the average number of cases was 51 per 100 000 population. In the republic as a whole by 1994, it had risen to 148.5 per 100 000, while in the critical Homel' Oblast, the total was 163.8.[31] Does this development perhaps indicate that observation and diagnosis have improved since Chernobyl? It has been noted, for example, that in 1993, for the first time in the post-accident period, Homel' Oblast experienced an increase of 200 doctors, and that middle-rank medical cadres had been sufficiently replenished. By 1996, predicted Kazakov, the doctor shortage would be resolved as 2200 doctors were being prepared annually at the republican level.[32] Diagnosis of sicknesses, particularly at an early stage, has undoubtedly improved. However, a self-sufficiency in medical cadres alone does not explain such a dramatic rise in illnesses of all types, including diabetes. Such developments are disturbing and traumatic for the families of those afflicted.

The impact of low-level radiation is still being widely debated. Some scientists consider that it has weakened the body's resistance to the degree that a form of Chernobyl AIDS is prevalent today, whereby the human organism is more susceptible to all kinds of disease. One source noted that in Belarus there had been a threefold rise in the general rate of sickness since Chernobyl. Although no specific diseases had been caused directly by the 'Chernobyl Factor,' she wrote, all the effects taken together influenced the immune system and weakened it, especially in children. Depending upon their age and individual characteristics, the radiation susceptibility of children was 9–21 times higher than that of adults. Children born in 1988, for example, to mothers who resided permanently in the regions of strict control, were weak, easily infected, and poorly developed compared to children from the clean zones.[33] Again, however, I must refrain from sweeping generalizations on the phenomenon of Chernobyl AIDS: there has clearly been a marked deterioration of children's health in the areas affected by Chernobyl, but one must take into account both non-radiation and radiation factors in assessing its origins.

Studies of the health status of children in the city of Minsk support the latter statement. Though a prophylactic study of 1990 revealed some 4218 'low-level radioactivity anomalies' unrelated to the Chernobyl accident, the dangers posed by such releases were far less serious than those that emanated from industrial pollution. Though the peak year for such pollution in the period 1980 to 1993 was reportedly the year 1985, the residents of Minsk were still being subjected to a formidable array of byproducts in the atmosphere, including formaldehyde, benzopyrene, and nitrogen dioxide. In the republic as a whole, 50 percent of city residents were being exposed in 1993 to toxic pollutants exceeding the maximum permissible norm by more than five times. The analysis of 1200 soil samples in Minsk indicated high concentrations of metals, and particularly lead, in virtually every part of the city with the exception of the west and northwest districts.[34] Any assessment of the impact of Chernobyl radiation must take into account such disarming figures when analyzing the health of the population.

To be sure, there are extreme interpretations on both sides of the Chernobyl spectrum. Since 1988–9 there have been wide divisions on health effects between 'scientific experts' and the general public and anti-nuclear elements in general. A Canadian physicist, Jovan Jovanovich of the University of Manitoba, describes the sentiments of the latter as follows:

> Scared and alarmed people started believing every health disorder and every unusual event happening in their environment was the result of the Chernobyl accident. Media were now reporting all these complaints and giving them a form of legitimacy. People believed the media and would get even more scared, more 'radiophobic', and would complain even more. Media would report these new complaints. ... Local and often republican authorities yielded to these pressures. ... The central government authorities, who for some reason were still holding back much detailed information, lost the confidence of people.[35]

Jovanovich may have a point, though he may have unwittingly focused on the heart of the problem in noting the actions of the central authorities in failing to provide accurate information to the public. Frequently, republican specialists – and it should be emphasized that one can often add the qualifier 'regional' specialists – have a tendency to attribute all medical problems to the radiation factor. As a phenomenon related to sickness, radiation is often given undue prominence. Initial misinformation from the Soviet authorities and the general incompetence of Belarusian state authorities on some matters compounded such beliefs. Kazakov has

pointed out that some people are inclined to believe that the entire catastrophe is a result only of the influence of radionuclides.[36] A comprehensive sociological survey conducted in late 1994 in the contaminated zones
revealed that 73 percent of those surveyed believed that their state of
health had deteriorated over the past 5–7 years.[37] On the other hand, some
sources continued to believe that the fallout of radionuclides had had no
impact thus far on the health of the population. According to Kazakov,
many foreign experts could be included in this group.[38] How would one
account for such skepticism, such a rift between a reportedly suffering
local Belarusian population and a skeptical group of doctors, scientists,
and medical specialists from the West?

One reflection of the latter attitude was the 1991 report by the
International Advisory Committee (IAC), established under the auspices
of the IAEA, based on a study of seven contaminated and six control villages,[39] which possessed a cesium content in the soil of less than 1 cu/km^2,
and the examination of the state of health of 1356 people born in different
years, both before and after the nuclear accident. The IAC drew up a
variety of conclusions and recommendations. It observed that there were
significant health disorders in both the contaminated and noncontaminated
regions, but none that could be attributed *directly* to exposure to radiation.
It criticized quite strongly the regional medical investigations, which had
been conducted at a low level and had often produced contradictory
results. It recommended also that because of the general fear of radiation,
a significant educational program was required for both teachers and local
doctors on the effects of radiation on the state of health.[40]

The IAC Report did acknowledge the significance of psychological
problems, noting that the overwhelming majority of those examined in the
affected regions 'believed or suspected' that they had a sickness that could
be linked to radiation.[41] It did not detect any sicknesses among children,
declaring that those examined were generally in good health. It acknowledged, however, that the survey was not detailed enough to preclude a rise
in some tumors in the future, particularly in the incidence of thyroid
cancers. Jovanovich, a member of the IAC, notes that the project could not
intervene in Soviet domestic affairs and that its survey was restricted by
the amount of funds and manpower available.[42] Its members were prepared to amend conclusions that in retrospect appeared to be erroneous.
According to Academician E. Kanaplya, the IAEA was open to his criticisms of the IAC Report in the following year, when he provided information that indicated the sort of problems being faced in Belarus.[43]

However restricted the scope of the IAC Report, there can be little
doubt that its impact was enormous. The fact that it was conducted under

the aegis of the IAEA, an organization that had been involved from the first in the aftermath of Chernobyl, and allowed an extremely privileged role by the Soviet government, gave it a certain authority. Its summary report was widely circulated among the media, which focused on the apparent negative health consequences of Chernobyl. In Belarus, Ukraine and other areas affected by Chernobyl there was a sense of moral outrage at the Report's conclusions. It was noted that the failure of the scientists to investigate the state of health of cleanup crews and evacuees rendered it an incomplete survey. Some regarded it as callous. Paradoxically, its research was conducted just before the most significant health problems in Belarus – thyroid cancers among children (see below) – came to light, though its circulation appeared to coincide with this new information. Arguably, then, the conducting of such medical research only four years after Chernobyl can be considered premature.[44] However, given the demand for such an inquiry, the concept of research on the health (and environmental) effects of Chernobyl can be justified. The problem lay rather in the limited nature of the research and the attempt to provide authoritative conclusions.[45]

In addition, in the areas affected by Chernobyl, there was an attitude of distrust toward official organizations that dealt with Chernobyl from the outset. A Chicago doctor has noted, for example, that the IAEA's reputation in the republics had been tarnished by its 'perceived role in promoting nuclear power and its ties to central Soviet authorities.' He cites a Ukrainian opposition leader's remark that: '[The IAEA report] is a script that has already been written. And if by some chance it is not, no one will believe it anyway.'[46] The fact remains that for a considerable period, the IAC Report remained the only serious attempt to address the impact of Chernobyl on health on a relatively comprehensive scale.

THE LIQUIDATORS

One group that did not receive attention from the team of international scientists was the cleanup workers – usually referred to by the Stalinist term of 'liquidators' – who were responsible for decontaminating the Chernobyl reactor building itself and the zone immediately surrounding it. Their numbers are difficult to estimate. In the former Soviet Union as a whole, a figure widely cited is 600 000 in total, encompassing the years of the aftermath of Chernobyl.[47] One source maintains that the number of Belarusian liquidators was 17 622.[48] According to more recent information, the figure was over 66 000.[49] All the cleanup workers received significant doses of radiation. According to one source, they exceeded the official lifetime limit of 35 rems for several days and even weeks. The majority of those involved '[did] not know their dose,' but many were sick.[50] An official report of 1994 in Belarus noted that the liquidators recorded the highest average doses of radiation received. Full body scans revealed that 30 percent of those examined received a dose of 50–100 mSv (5–10 rems), 47 percent received 100–250 mSv and 7.3 percent 250–500 mSv. Approximately the same doses were found on those evacuated from the 30-kilometer zone.

In the period 1987–91, the liquidators suffered from a significant rise of illnesses of the nervous system; psychic disturbances, diseases of the blood and blood circulation organs, sicknesses of the digestive system and the nebulous 'vegeto-vascular dystonia' (discussed below). Studies of the cleanup workers at the Kirov military medical academy gave rise to several conclusions about their state of health. Similarly disturbing results were manifested in the following years. Over the period 1991–4, the morbidity rate associated with disorders of the hematological system, respiratory organs and the digestive system among the liquidators was higher than the republican average for adults. How much higher? The incidence of hypertension was said to be 7.5 times higher than the national average; stenocardia 3.3 times higher, bronchial asthma 2.2, bronchitis 5.8, and digestive disorders 10–12 times higher than average.[51] Concerning diseases of the thyroid, incidence among 'liquidators' was reported to be 40 times higher than the republican average.[52] The morbidity level among liquidators from all regions was reported to be especially high (36 000 people).[53]

Early in 1993, information on liquidators was published in a medical journal by Ukrainian experts. Scientists had studied 426 men aged 21–45 who participated in decontamination work. Among them, about 9 percent were said to be suffering from acute radiation sickness of the first to third stage, while among the remainder had been observed the phenomenon of 'vegeto-vascular dystonia.'[54] The study focused on the sexual activity of the men prior to the disaster, and in the first months and years after it. It established that 38 percent of these liquidators, many of whom were under the age of 40, were suffering from disorders of the sexual functions. They complained of a loss of libido, and in some cases its complete absence. Forty-five percent declared premature ejaculation during intercourse and a reduction of sperm production. Clearly factors not related to radiation also need to be taken into consideration in analyzing this report. Many liquidators were said to be suffering from alcohol abuse, while others displayed symptoms of hypochondria, ultra sensitivity and irritability. The doctors, it was noted, could only concentrate on the original sickness, such as radiation sickness or 'vegeto-vascular dystonia' in the hope that both sexual problems and those in the psychological and emotional sphere might be eliminated.[55]

The total number of liquidators who have died since the accident is not known, though various figures have been issued. A source in Ukraine, a country in which the Chernobyl Union of Liquidators has regularly issued figures on mortality rates, claimed in late 1994 that in Belarus one in six of those who have died committed suicide. This same source stated that in Ukraine over the eight post-accident years, 93 000 people living in the contaminated zones had died.[56] There was no indication how many of these people were liquidators or even how such a figure was ascertained. Earlier figures were cited by Volodymyr Chernousenko, who worked in the 30-kilometer zone around the reactor, and declared that between 7000 and 10 000 liquidators had died as a result of Chernobyl.[57] Two points should be made about such figures. The first is that they are hardly unrealistic if one considers the number of people involved in the decontamination campaign. Certainly it can be demonstrated that the health of the liquidators has declined significantly since 1986. Second, such figures have not been corroborated by independent sources. To my knowledge no complete list has ever been published with the names of the alleged victims, how they died, their ages and previous medical condition, or even their location. Without such a comprehensive register, one cannot accept such figures at face value. In some ways they detract from the discernible health effects of Chernobyl because they suggest inflation of figures or the use of figures that cannot be determined definitively. More reliable figures are those compiled by regional authorities that kept records of personnel – usually militiamen – who were obliged to serve in the zone. In Vitsebsk region, for example, 19 'liquidators' who served at Chernobyl during the immediate aftermath of the accident have died to date, and this figure is corroborated by several sources.[58] An article published on the ninth anniversary of Chernobyl in Belarus confirmed a current total of 125 deaths among those liquidators resident in Belarus.[59] The total suggests that the casualty rate to date is likely to be high, but the precise figure is difficult and may even be impossible to ascertain.

PSYCHOLOGICAL PROBLEMS

As noted above, psychological problems are difficult to discern and evaluate. As one scientist had observed, the subject of radiation medicine psychology did not even exist in the Soviet Union at the time of the Chernobyl accident.[60] However, after the initial shock of the disaster in 1986, several prominent figures in the medical establishment used the term 'radiophobia' to account for the emotional and sometimes irrational behavior of the population. The term was often misused and in some respects it was used to denigrate local reactions and to explain local outrage at the failure of the Soviet authorities to respond appropriately to the new conditions, especially after the spring of 1989 when radiation maps had been published in *Pravda* and elsewhere. The use of the term 'radiophobia' did not, however, indicate that officials were delving deeply into the psychological impact of Chernobyl. Only by 1990 were several investigations of the phenomenon and of psychic tension generally in the affected regions under way.

One of these surveyed 500 residents of areas of high contamination, noting that tense psycho-social situations and anxiety were pervasive. The population had reportedly changed its lifestyle. Over 60 percent no longer carried out the traditional gathering of mushrooms and berries in the forests; 50 percent had stopped swimming and sunbathing; 39 percent felt general discomfort; and over 52 percent planned to relocate. If this sort of reaction constituted radiophobia, then what was its cause? First, it was declared, radiophobia arose from a lack of knowledge of the rudiments of radiation sanitary measures. Second, the local population distrusted official information. Third, it overestimated the effects of low radiation doses on health status. Finally, the population's fear emanated partly from the inadequate conduct of medical personnel and officials in the contaminated zones.[61] A similar survey, which encompassed 500 residents and 50 doctors and nurses in the most contaminated areas, examined the causes of radiophobia and confidence in different sources of information. The leading causes of radiophobia were declared to be:

1. Delayed official information (78 percent of those surveyed).
2. Contradictory information from different official sources (61 percent).
3. Absence of official information (54 percent).

Even the medical staff lacked sufficient information about the radiation situation in their region, while the population placed its faith mainly in lectures by experts.[62]

Two years later, the situation in the republic appeared to be more serious. After 1991, there was no longer such a desire to leave the contaminated areas. The change was abrupt and is not easy to explain. Only 11 percent of those who were eligible for evacuation wished to move residences. However, a distinct generational difference was evident.

Reflecting a general demographic trend that is unrelated to radiation effects, the young people generally left the villages and the old remained behind. Perhaps by 1991, all those who wished to move or be moved had left the area, leaving behind the elderly with their commitment to their place of birth. The working environment, however, was deteriorating, especially sanitary conditions. Those remaining behind expressed a helplessness to resolve their problems; they noted the apparent indifference of the authorities, and their future appeared to be grim. Such feelings generated stress. What made matters worse was the departure of many doctors and teachers.[63] By the spring of 1993, the Homel' branch of the Institute of Radiation Medicine reported that 70 percent of the population living in the zone of acute control had suffered diverse psychic violations.[64] Its central branch in Minsk also offered some disturbing figures. Its analysis of the mental state of children and teenagers indicated the formation of 'mental disadaptation' and neuropsychic disorders. Among children living in the contaminated zone as compared to clean zones, the frequency of hypochondria was four to five times higher, phobia three times higher, and asthenia twice the levels.[65]

In the summer of 1993, a conference was initiated in the town of Mazyr, Homel' Oblast, on the initiative of the State Chernobyl Committee, the Belarusian Ministry of Education, the Mazyr Pedagogical Institute, and the Center of Chernobyl Psychology and Pedagogical Problems at the Belarusian National Institute of Education. The focus of the conference was the problems of socio-psychological rehabilitation, and the social and juridical protection of children and adolescent victims of Chernobyl. As the first major Belarusian conference on the psychological impact of Chernobyl, its conclusions merit discussion in detail.[66]

The article concerned with the conference included a personal assessment by its author, who revealed that today (i.e. the summer of 1993), for most people, the problems associated with Chernobyl had shifted to the background. Other concerns were paramount, particularly the difficult economic situation, and the difficulties of feeding and clothing a family. People, it was stated, wanted peace, a normal life. They were anxious to forget the tribulations of a disaster. Mankind, it added, can become accustomed to the worst events, even radiation and terrible prognoses about future health, and especially when the worst predictions 'do not come true.' Nevertheless, the preoccupation with daily problems did not signify that the disaster had disappeared as a source of tension. The Chernobyl victims merely pushed such thoughts to the background, where they 'gnawed away.' It was such thoughts, in the view of the writer, that had led to deviations in behavior: aggression, conflicts, and distortions in psychiatric development, particularly in children. The victims were being well treated, but when children became sick, many of the diseases were linked directly or indirectly to Chernobyl. Health has several aspects, noted the author. If we wish our children to be healthy in every sense of the word, then psychological health is as important if not more important than physical health. In the post-Chernobyl situation, however, children were least protected in this sphere. Even those born after 1986 had received a 'negative emotional exposure.' Children do not grow up in isolation, the author stated, and their psychic development was being influenced by teachers and the anxious family atmosphere in many regions.

At the conference, an unnamed specialist from the World Health Organization (WHO) was cited as stating that deviations in personality development could not be attributed to radiation exposure, especially when the morbidity level was unknown. Probably therefore they were a result of psychological factors and stress. Yet the fact that they were being attributed to the impact of radiation not only promoted the increase of psychological strain and provoked additional stress influence on the health of people, but also undermined the population's faith in the competence of specialists engaged in radiation protection. Evidently similar conclusions were expressed by the Red Cross and other conference delegates. The adverse psychological conditions were exacerbated by economic problems, creating painful educational dilemmas, especially for schools in the contaminated zone. In general there was perceived a critical shortage of teachers for such schools. In the town of Dobrush (Homel' Oblast), for example, first-grade pupils at the new gymnasium had been monitored, and a serious reluctance to study and a lack of interest in acquiring knowledge were noted. The majority of adolescents also had no confidence in their future, and were generally characterized by pessimism, apathy, and a careless attitude toward their studies.

Psychologists who had observed children in the contaminated zone, it was noted, considered that Chernobyl represents a powerful psychological trauma for the population, and especially for children. In such an unfavorable background, children grow up with the firm belief that they will not be healthy, or obtain a good job. They do not foresee themselves entering university, or starting a family and having healthy children. All efforts of teachers to instill in them a more optimistic attitude, it was conjectured, would be in vain unless the psychological aspects of the Chernobyl health problems received more attention. Psychiatrists in Belarus, however, were still associated with an image of the 'madhouse.'

Finally, a study conducted under the auspices of the Belarusian State University examined secondary school students during the pre- and post-Chernobyl periods, based on five groups of students, 286 in total, aged 11–12.

1. A control group of students from the Homel' region, whose state of health had been examined in 1985;
2. Students from the same schools screened in 1991;
3. Students from Vetka, where contamination levels of the soil by radioactive cesium were between 1 and 5 cu/km^2;
4. Evacuees from the contaminated zone now living in the city of Minsk;
5. Students who have resided permanently in Minsk.

The investigation looked at the condition of the nervous system. It measured also visual, acoustic and motor parameters, the cardiovascular system and other psychological functions. Students were also given basic mathematics tests, and tables of letters and numbers to test their concentration. The researchers concluded that those students living in highly contaminated zones (15–40 cu/km^2) had suffered losses in short-term memory capacity and speed of thought, even for lengthy periods after resettlement. Evacuation, nevertheless, reportedly enhanced the ability to concentrate, which was said to be particularly poor in the contaminated regions. Resettlers also suffered from more problems of the nervous system and what was termed 'working stress.'[67]

Clearly such attempts to provide an analysis of the psychological effect of a disaster on the population are in their infancy. That the population of Belarus generally has been affected psychologically is apparent to any observer visiting either the city of Minsk or the contaminated zone. It has been pointed out, however, that no consensus exists in works on the psychological impact of Chernobyl. Some scientists, for example, link the psychiatric problems directly with the degree of additional radiation received; others consider that the term 'radiophobia' can be applied to encompass all such problems.[68] One can say at the least, that any attempt to provide a definitive account of the health effects of Chernobyl must include the psychological impact, however complex the topic may be. As noted earlier, it was the factor most evident to the scientists compiling the IAC Report at the behest of the IAEA. Further, the enforced relegation to the background of problems associated with Chernobyl has evidently served to increase tension, to foster anxiety and suspicion about health symptoms.

LEUKEMIA

One of the biggest fears concerning the impact of Chernobyl was that the significant rise in radiation background would result in an increase of leukemia among the affected population. Leukemia appears before other types of cancer. According to one specialist, it starts 2–3 years after exposure, and peaks at 6–8 years.[69] Another expert notes that at Hiroshima and Nagasaki, after the dropping of the atomic bombs, the first reports concerning a rise in the number of leukemia cases appeared after only 18 months, statistical data were available after 3–4 years, and the peak of the increase in cases occurred at 5–6 years.[70] Thus medical specialists anticipated that an increased incidence of leukemia would become the first discernible consequence of increased radiation levels in the atmosphere and the soil. In the Republic of Belarus, however, prior to 1988, there was no such scientific sphere as epidemiology and especially hematology. It had to be created from scratch, with the necessary training of medical personnel. By 1992, an infrastructure had been developed with a total of 200 people working in this area, but according to the director of the Institute of Hematology, the lack of knowledge about leukemia in the republic had resulted in the wildest speculations about its appearance. Thus in Naroulya (Homel' Oblast), it was rumored that of 24 children in one settlement, 12 had died of leukemia, even though over the previous four years not a single case of the disease had been uncovered.[71] It is clear therefore that Belarus, the republic most affected by radiation from Chernobyl, was unprepared for the monitoring or treatment of leukemia.

The subject is closely associated with the name of Dr. Evgeniy Ivanov. Though the ostensible pioneer of the Institute of Hematology in Belarus, and the leading authority on the question of leukemia in the republic, he is a controversial figure whose conclusions have aroused much ire both among his peers and the community at large. Nevertheless, he has dominated the scientific literature on the subject to such a degree that it is necessary first to analyze closely the findings of Ivanov, then outline the criticisms of his opponents, and only then to attempt to reach a conclusion on the occurrences of leukemia in the republic as a result of Chernobyl.

Ivanov's findings have been published *inter alia* in two major articles in the conservative weekly *Sem' dney*, which ensured mass circulation of his conclusions. The first article began with a preamble that the statements therein represented the results of six years of study at the Institute of Hematology and Diseases of the Blood in an attempt to create the first Register of Diseases of the Blood of Children and Adults in the CIS countries. The study had embraced not only local experts, but also foreign specialists from the United States, Switzerland, France, Japan and Britain who had visited the Institute in the period 1989–92. Its conclusion was that no rise in leukemia had been noted in the republic, and that the existing rates – 40–48 children per million population – were well within the European average range. Not a single case of leukemia had been detected, claimed Ivanov, which could be attributed to the Chernobyl radiation factor with any certainty. A detailed chart compared cases of leukemia among children aged 0–14 years in Europe between the years 1980–5, and Belarus in 1979–91 in terms of sicknesses per 100 000 children, and divided the subjects into males and females. The chart demonstrated that the incidence of leukemia in Belarus was clearly within the European average, though perhaps as high as the 75th percentile. The highest incidence was recorded among males in the Czech Republic, followed by males in Denmark, Belarus and Austria at between 47 and 52 cases per million children. Notably also, the rate of leukemia among children had risen significantly in Belarus between the periods 1979–85 and 1986–91.[72]

Results of the examination of children up to 14 years of age only in Belarus in the post-Chernobyl years are summarized in Table 4.1.

Ivanov also demonstrated another theory; that the highest incidence of leukemia had occurred in the areas of the country that suffer most from industrial pollution, particularly the city of Minsk, in which levels exceeded those in Homel' and Mahileu. He has noted elsewhere that the use of five-year periods can also be misleading. The peak year for incidence in Belarus was 1987, when levels reached 60 per million among boys, and 30 among girls, at a time when it was decidedly too early to detect results from Chernobyl. In Homel' region, in which an increase of

Table 4.1   *Incidence of leukemia in Belarus, 1979–91 (per million children)*

Oblast	1979–85	1986–91	Total cases
Brest	41	40	196
Vitsebsk	39	42	145
Homel'	35	40	192
Hrodna	36	38	132
Minsk	35	40	190
Mahileu	48	41	174
City of Minsk	51	48	218
Overall	40.7	41.3	1247

Source: Sem' dney, 16 January 1993.

leukemia had been recorded, the peak year was 1986.[73] Though admittedly, he says, it is easy to make errors in assessment, it is clear that radiation has not affected the incidence of leukemia.

Further elaboration of Ivanov's conclusion was made available in the
summer of 1993. By this time, the republic had been included in two inter-
national programs for the study of children's leukemia and lymphoma: the
International Program on the Medical Consequences of the Chernobyl
Accident under the auspices of the WHO's 'Hematology' program, based
in Switzerland; and the Research of Children's Leukemia and
Lymphoma's European program, based in France. Ivanov's statements in
July 1993 were reportedly based on a conference at the Institute of
Hematology at which all the principal medical specialists on the subject
were in attendance. Thus at the outset he wished to make plain that the
conclusions expressed were not his alone, but rather the consensus of his
own specialists backed by those of the WHO and other international
groups. In the year 1992, 104 Belarusian children had become sick with
leukemia. The figure was disturbing, but was lower than in 1991 (108
cases) and 1990 (111 cases). It was considerably lower than the peak pre-
Chernobyl year of 1979 (123 cases). But how did this total compare with
figures in Europe and the United States? Ivanov noted that the world
average for children's leukemia was between 35 and 60 cases per million
children, and Belarus fell within this range.

In the period 1979–92, the incidence of children's leukemia, according
to Ivanov, had not changed within Europe or the United States as a result
of the Chernobyl influence. Hematologists in Russia and Ukraine had
reportedly received similar data to that in Belarus. In 1992, the highest
level of the illness was in Homel' Oblast, with 59 cases per million chil-

Table 4.2 *Highest recorded levels of children's leukemia (per million
population) 1979–88*

Oblast	Year	Level
Minsk	1979	72.5
Minsk	1981	71.0
Hrodna	1979	68.0
Vitsebsk	1988	66.5
Vitsebsk	1979	60.0

Source: Ivanov, 3 July 1993.

dren. However, Homel' was followed by less contaminated regions: Hrodna with 52.2 cases, and Vitsebsk, with 52.0 cases. In Homel' Oblast, over 80 percent of those sick were from the city of Homel' and the contaminated areas. Yet Ivanov was not disturbed by the figures. He pointed out that in other oblasts prior to Chernobyl, the incidence rate was higher than the 1992 figure for Homel'.

Hematologists, declared Ivanov, had critically examined mistaken figures about a rise of leukemia among children in Belarus that had been released in 1989–90 and reached a different conclusion: that after six years, the radiation factor had yet to manifest itself as a factor in this type of illness. Yet children's leukemia as a disease was still making 'slow but constant progress' worldwide. So what factors were behind its incidence?

Here, Ivanov indulged in some dubious data massaging. In the period 1975–85, he noted, 43 out of a million children worldwide contracted leukemia, whereas in the period 1986–92, the figure was 45, i.e. a rise of two cases per million. In Belarus, the disease occurred in both clean and contaminated zones. And even if one maintained that the additional two cases might be connected with radiation, then the other 40 must derive from other factors. Hence in the republic there existed a factor or factors that were 20 times more influential than radiation. What might this factor or these factors be? Ivanov was unequivocal: the chemical pollution of Belarusian cities was the likely cause for the general rise in children's leukemia. Emissions from auto transport in the city of Minsk alone composed over 125 000 tons of harmful substances each year, while industrial enterprises released byproducts such as sulfuric acid, phenol, pesticides and other products. Such a situation would help to explain the high incidence of children's leukemia in the city of Minsk.[74] Paradoxically, Ivanov does much to discredit his case by such a simplistic analysis of the figures he has produced. Before looking at them in more detail, let us turn to a critique of his article which appeared in the same source in July 1993.

Leonid Karkanitsa, a Candidate of Medical Sciences, provided an exhaustive response to Ivanov which, at the least, cast considerable doubts upon the Chief Hematologist's contention that leading specialists in the republic agreed unanimously with the views expressed above. Karkanitsa's point of view, it was stated, was quite different from that of Ivanov. When discussing people's health, it was important to introduce all aspects of a problem without omissions, he added. Above all, he expressed disagreement with the results produced. In the first place, Ivanov's views did not represent those of the 'collective' of the Institute of Hematology and Diseases of the Blood. Many scientists and medical specialists had indeed participated in the gathering of data, but at the stage of elaboration

– and analysis in particular – access to this information had been restricted to a small group. The body of the work had been conducted in the second half of 1989 as a continuation of the former all-Union program, under the supervision of the Ministry of Health. Only in 1993 had the WHO begun to supervise the program and even then not with a great deal of investment. It is not correct to declare, as does Ivanov, that the research on children's leukemia was conducted with the expertise of foreign experts. maintained Karkanitsa. Certainly such people were permitted to observe certain data, but this amounted to a discussion over a cup of tea, and not 'expertise.'[75]

Concerning the radiation factor, Karkanitsa asked what exactly was known. First, radiation was the chief component in the influence of short-living isotopes (iodine and others) in April–May 1986, and after their disintegration, of the possible influence of long-living isotopes, particularly strontium and cesium. From the fallout scientists have deduced only the approximate so-called collective doses of irradiation of residents in certain areas – far from all of them – and the level of contamination of a given territory. The latter is also a complex matter because of the unreliable oscillation of levels even within the boundaries of a single population point. Lost irrevocably in these calculations was the most important factor: the individual radiation dose. In brief, then, noted Karkanitsa, the radiation factor had not received its full characteristics. The role of this factor was impossible to underestimate and it was therefore facile to study it only at the level of general contamination, without calculating the individual dose, the migration of the population, demographic changes, and other factors. Ivanov himself had acknowledged that the previous, i.e. Soviet, leadership had not sought to provide reliable information. And even if the authors of a thesis on leukemia among children had somehow managed to collect all available information, all cases of leukemia commencing in 1979, it would still be unreliable. Their means of diagnosis were inadequate and children's hematology generally remained at an unsophisticated level. In the research itself, he noted, it was impossible to establish the relationship of phenomena (leukemia) and the causal factor (radiation) with such gaps and omissions in information. In order to make a definitive conclusion every case of leukemia would have to be characterized fully, with supporting diagnoses and the registration of each individual dose of radiation.

Ivanov's conclusions were not supported by such data. Moreover, he had introduced only quantitative results of leukemia sickness. Did not qualitative variations merit a place also? No one would deny, declared Karkanitsa, that chemical or general contamination of the surrounding environment is extremely unfavorable for the health of the population, but one must also apply the laws of science. The level of contamination as a causal factor in each concrete case would have to be determined, and then applied to a large number of cases to establish causality and reliability in order to affirm or not to affirm that industrial pollution was making a person sick, and to evaluate pollution as a factor alongside other risk factors. Otherwise, such conclusions must be considered unfounded. The city of Minsk, for example, contained the most leukemia victims and also suffered from the highest industrial pollution levels. Therefore, Ivanov had declared, pollution was the chief cause of leukemia. However, this conclusion appeared to ignore the fact that families from the contaminated zone had resettled in Minsk over the past few years.[76] But if one agrees with the author's argument concerning Minsk, why would it not apply to cities that are even more heavily polluted than Minsk, asked Karkanitsa, namely Mahileu, Hrodna and Novapalatsk? Such arguments at best could be regarded as working hypotheses. In an assessment of the causes of children's leukemia, which after all is a worldwide phenomenon and occurs even in the ecologically cleaner Western countries, one needed to take into account other factors such as genetically predetermined cases.

Karkanitsa's arguments have been cited here in full because they point out some serious flaws in the statements of Ivanov, and the latter have been widely circulated in the form of articles and media interviews. The most that can be declared with accuracy is that to date there has not been a discernible sharp rise in children's leukemia that can be attributed with certainty to the effects of radiation. Statistics provided by Ivanov do indicate, however, two factors that appear to contradict the previous statement: first, a rise in leukemia generally in Belarus; second, a rise in which the highest level of increase includes those areas most highly contaminated, i.e. in Homel' Oblast. In terms of incidence per million children, the republic is within the world average for the disease, but it has risen to the very upper end of that average. To reiterate, however, although Ivanov's methods of making deductions appear highly irrational, there does not appear to be any reason to doubt the authenticity of his figures: there has been no appreciable increase of children's leukemia in Belarus and given the time lag since Chernobyl, there is unlikely to be such an increase in the future.[77] There is, however, a rise in the number of chronic – as opposed to acute – cases of leukemia. The increase has been declared a cause for serious concern, though detailed results were lacking in the period 1992–4.[78]

One may conclude this section on leukemia by stating that notwithstanding the limited nature of the inquiry and what appears to be inflated optimism on the part of the Chief Hematologist of Belarus, the health consequences of Chernobyl in terms of incidence of child leukemia appear to be surprisingly limited, even though the link between the disease and increased radiation is evident.

THYROID CANCER AND OTHER DISEASES OF THE THYROID

The most significant and worrisome medical impact from Chernobyl in the republic is the marked rise of thyroid cancers among children in the contaminated regions. The rise was not noted in the IAC Report (it had not occurred among those children investigated), and the connection between thyroid cancer and the nuclear disaster was denied by Leonid Il'in, the head of the Soviet health authorities examining the effects of Chernobyl. Il'in declared during a visit to Belarus in 1990 that such cases were a result not of Chernobyl, but solely of a known iodine deficiency in the soil.[79] Leaders of the Ukrainian Scientific Center of Radiation Medicine anticipated an eventual rise in thyroid cancers in regions affected by Chernobyl in Ukraine of 300 cases (30 incurable) over the next 35 years, or a rise of 1.4 over spontaneous thyroid morbidity for children.[80] What was unexpected was that the increase would occur so quickly and on such a dramatic scale, particularly in Belarus. Thyroid cancer has become the most discernible health impact of Chernobyl and a cause for serious concern in the republic. Because of the controversial nature of these findings, let us begin this analysis with some Western and Japanese reactions to the phenomenon before looking in depth at the investigation and reaction in Belarus.

In September 1992, several medical doctors put their signatures to a letter that appeared in the British journal *Nature*, concerning the outbreaks of thyroid cancer in Belarus. The doctors, who were from Switzerland, Italy and Britain, had visited the republic under the auspices of the WHO Regional Office for Europe and the Swiss government to examine the children and study the cases. They were given access to and allowed to peruse the full medical records of 11 children who had had recent surgery and were recuperating in hospital. All the children had been diagnosed in the period 1989–92: eight of them had lived in the Homel' region prior to the accident, and two in Brest Oblast (the eleventh child lived outside the contaminated zone). Their ages ranged from four to 13, with the youngest being born two days after the accident.[81]

In all but two cases the WHO delegation agreed with the local diagnosis of the children. Data on the incidence of children's thyroid cancer in

Belarus revealed that it had undergone a marked increase in frequency from 1990 onwards, which the visitors found to be a much shorter time than the average for cases of exposure of infants to external radiation. Though the eight youngest children were at the fetal stage of development when exposed to radioiodine, the fetal thyroid begins to concentrate iodine at 12–14 weeks. The tumors were noted to be aggressive with a high proportion of malignant nodules. The cases of thyroid cancer had risen much more rapidly in children than in adults, and had greatly exceeded the average republican incidence of this disease in children under the age of 15, which was about one per million per year. In Homel' region with a population of 2.5 million,[82] levels were already 80 times this average. The cause of these cancers was the exposure of the neck to radiation. The authors declared that both in the level of exposure and the number of people exposed in such a brief period, Chernobyl was a unique accident that does not bear comparison with other cases of high radiation exposure: the Marshall Islands; the Windscale nuclear accident in Britain; weapons fallout; and nuclear accidents in general. Even in Japan, after the atomic bomb was dropped on Hiroshima in August 1945, where scientists recorded a close link between the radiation dose and outbreaks of thyroid cancer, the radiation received was mainly from external exposure and the fallout results unclear.[83]

In Belarus, however, there appeared to be a strong case to make more definite conclusions. The doctors maintained that:

> Experience in Belarus suggests that the consequences to the human thyroid, especially in fetuses and young children, of carcinogenic effects of radioactive fallout, is much greater than previously thought.[84]

In short, a direct correlation was perceived between exposure to high levels of radiation and the development of thyroid cancer among children in the republic. The authors make it plain that such a development was unexpected, at least in so short a time and to such a degree. Their conclusions were quickly assailed by more skeptical analysts, who maintained that the data used were too limited to reach such an unequivocal conclusion. Two doctors writing in the same issue of *Nature* took issue with the authors' findings, maintaining that there were as yet insufficient data to establish a definite link between radiation exposure and the increase in thyroid cancers, and that information on the actual radiation dose of the children was critical.[85] One of these doctors, incidentally, was the chairman of the International Advisory Committee which compiled the International Chernobyl Report. Since the new findings undermined the validity of that report, a critique from such a quarter was hardly surprising.

In an earlier publication, several Belarusian medical specialists (including, incidentally, Ivanov) noted that the rise in thyroid cancers in the republic was a cause for special concern. In the post-accident years 1986–90, thyroid gland cancer (TGC) among children had risen by seven times in the republic – calculated per 100 000 child population – and by 19 times in Homel' Oblast. In 1986–90, according to data provided by the Thyroid Tumor Clinic in Minsk and the Institute of Radiation Medicine, 45 TGC cases had been registered as compared to only seven cases in the analogous period in 1981–5. In the period January–May 1991, the Thyroid Tumor Clinic performed 20 operations to remove thyroid tumors. Fourteen of the children subject to surgery were from Homel' Oblast, two from Vitsebsk, one from Minsk Oblast, and three from the city of Minsk. Altogether in the period January 1990 to May 1991 doctors had diagnosed 51 cases of TGC among children, of which 29 (57 percent of the total) were in Homel' Oblast, and among whom at the time of the accident 60 percent were aged 0–5 years. Only 18 of the 51 had been discovered at the beginning stage of the tumor.[86]

Not all the cases represented a clear-cut correlation with the Chernobyl fallout. An outbreak of thyroid cancers among children in the Svetlahorsk region puzzled observers. According to M.A. Krysenka, Deputy Minister of Health, 190 children with thyroid cancer were registered in May 1993. Of this number, seven were from Svetlahorsk, a fairly 'clean' town, but the location with the highest rate of morbidity in Homel' Oblast. Krylenka points out that thyroid cancer was first discovered among children of the region in 1988. In the settlement of Shatsilki, Svetlahorsk Raion, one girl eventually died of complications arising from the original thyroid cancer after the disease had progressed rapidly. Prior to that date, goiter condition had been detected as an illness among adults. Homel' Oblast in general is a center of endemic goiter[87] because the soil does not contain sufficient iodine. Thus for many residents, a diet rich in iodine had been prescribed: nuts, seaweed, sea fish and the like. After Chernobyl, noted Krylenka, once the fallout of radioactive iodine was detected, medical research was conducted more regularly. In 1988, 83 children were revealed to have pathology of the thyroid gland; in 1989, 807; and in 1990, 9924. Autonomous thyroiditis was also revealed.[88]

When children of Homel' Oblast were studied with ultrasonic devices, it was discovered that some of them had diminished gland segments or were completely lacking in one gland. Diminished glands, stated the deputy health minister, are unable to produce the necessary hormones and there is no available cure other than certain medications. In the summer of 1993, 76 children had been examined at the Institute of Radiation

Medicine in Minsk. Five were revealed to have signs of cancer and one case was confirmed. Twenty-eight had auto-immune thyroiditis and 13 had 'knotting formations' of the thyroid gland. Surgical operations had been carried out in three cases. These statistics were thought to be cause for serious anxiety.[89] However, they do not indicate that exposure to radiation was the primary cause of the illnesses. Nevertheless, evidence of such a link became increasingly apparent from a number of studies by medical specialists of the republic.

A report published in 1992 noted that at the end of 1990, 84 200 children with goiter at the first stage were observed in the republic; 63 150 with goiter at the second stage; and 601 at the third stage. The percentage of children with second-stage goiter varied from 16–20 percent in the Hrodna and Vitsebsk regions to 48–67 percent in the Homel' Oblast. Reportedly the number of cases was highest in the contaminated zones. The high percentage of children with hyperplasia of the thyroid gland, which had increased in recent years, was attributed both to the natural insufficiency of iodine in the southern regions of the republic, and the compensatory saturation of the thyroid gland with radioactive iodine in 1986. In comparison with 1985, it was stated, the number of children with pure goiter (beginning with the third stage) had risen by 300 percent. Prior to 1986 cases of thyroiditis were rarely diagnosed, partly because of a lack of adequate equipment for such a task.[90]

In September 1992, three medical specialists – Kazakov, E.P. Demidchik, head of the Thyroid Tumor Clinic, and L.N. Astakhova of the Institute of Radiation Medicine – published a paper that reiterated and expanded upon the revelations made in *Nature*. They pointed out that the rise in thyroid cancers among children in Belarus began in 1990 and has continued. The overall rise had started from an average of four cases per year from 1986 to 1989 inclusive, to 55 in 1991, and a projected 60 cases in 1992. This increase was not distributed uniformly across the country. No significant increases had been detected in Mahileu Oblast, the city of Minsk, or Vitsebsk Oblast. By far the biggest increase was in Homel' Oblast, where incidence had risen from one to two cases per year to 38 in 1991, with detectable but smaller rises in Brest and Hrodna oblasts. They pointed out that Homel' suffered from very high levels of radioactive fallout. The radioactive plume passed first over Homel' region in the first few hours after the major release of radioactivity, then moved over Brest and Hrodna regions. This initial fallout contained large amounts of iodine-131 and other short-lived iodine isotopes. The tumors detected had been classified according to WHO categories. Of the 131 cases, 128 were papillary carcinomas.[91] They were relatively aggressive tumors: 55 of the 131 cases indicated direct extension of tissues around the thyroid and six had metastasized, mostly into the lungs. Of the tumors, 23 percent were less than one centimeter in diameter. One of the children had died (as noted above) and 10 others were seriously ill.[92] The findings of the three Belarusian specialists did not differ appreciably from those of the WHO team cited above. Before long, more data were available. The most detailed and fundamental research into this illness has been conducted by Demidchik,[93] a man who has worked closely with his international associates in the field of thyroid cancer and one who has been notably open about sharing the results of his research. The following section, devoted to Demidchik's research, is based on two personal interviews, an unpublished paper, and materials faxed to me on Demidchik's behalf in January 1995.

DEMIDCHIK'S FINDINGS

In an April 1993 interview, Demidchik stated that his work at the Thyroid Cancer Clinic had begun in 1966, at which time thyroid gland cancer (TGC) was a rare disease in Belarus. There were at that time some 40 cases in the republic among both adults and children. Since then the rate had risen gradually but at a rate slower than that in other European countries: about one case per 100 000 inhabitants per year. Among children prior to Chernobyl, there were seven cases of TGC in the republic. Between 1986 and 1989 a small increase in TGC was detected among children: two cases in 1986; four in 1987; and five in 1988. However, in 1990, 29 cases were suddenly detected. By 1991, the figure had jumped to 59, or 15.3 percent of total cases. There was no corresponding dramatic rise of TGC among adults. Demidchik was unequivocal: using a map of the fallout of radioactive iodine in Belarus on 10 May 1986, he demonstrated that although some 75 percent of the republic was encompassed by the fallout, the hardest hit areas were Homel' and Brest regions.[94] A related study has indicated that the highest thyroid gland radiation accumulation was incurred by the residents of Khoiniki Raion, Homel' Oblast, who were subsequently resettled.[95]

Almost all the children with TGC in the republic were either born or conceived prior to the Chernobyl disaster. In 1993, the illness pertained to children born before the catastrophe and six who were at the fetal stage at this time. The ages of the children varied. Thirty children were either under one year of age, or slightly over one year. Thus even at such an early age, they were irradiated through their mothers' milk. Six children born during the intensive period of Chernobyl developed thyroid cancer within three to four years. Thus, concluded Demidchik, there could be no grounds for anyone to assert that the growth of TGC was unconnected with the Chernobyl disaster. In fact, in contrast to the situation with leukemia, monitoring of TGC before the Chernobyl disaster had been careful and thorough: possibly two to three cases at most may not have been detected. At this time, Demidchik anticipated some 800–900 cases in Belarus in the near future.

In an unpublished paper of late 1993, Demidchik observed that in the period 1979–92, those sick with thyroid gland cancer (TGC) – both adults and children – had been divided into two groups of identical seven-year periods: 1979–85, and 1986–92. After Chernobyl, 1097 new cases of this illness had occurred. In Belarus as a whole, and in all regions of the republic, there had been detected an increase in cancer of the thyroid gland of two times or more; but this rise had been particularly marked in children. The highest peak of TGC was found in those over 50, i.e. a period of life in which the functional activity of the thyroid is reduced, and cells are more likely to swell. By 1992, TGC among children had risen to 2.77 cases per 100 000 in the republic, and stood as high as 8.8 in Homel' Oblast. The first notable increase occurred in 1990. Thus, in Demidchik's view, the 18 children sick with TGC in 1986–9 did not derive the illness from radiation. However, concerning the 154 children sick over the period 1989–92, the inducement of TGC by radioiodine was 'fully possible.'[96] But how closely can one correlate cancer incidence with radiation fallout? Are there not likely to be significant exceptions, as in the Svetlahorsk case noted above? Is it simplistic to attribute the sudden rise directly to the influence of Chernobyl radiation? Might there be subsidiary or secondary causes that would better explain the remarkable phenomenon of a dramatic surge in TGC only four years after a disaster?

In January 1995, Demidchik presented new statistics at the Minsk Medical Institute, in Belgium and Germany. These allow the construction of the most detailed table to date: see Table 4.3.

Thus 53.8 percent of all TGC cases today originated in Homel' Oblast and 22.8 percent in Brest Oblast. The number of cases in the republic has continued to rise at a substantial rate each year and there is no indication that the illness has yet reached a culmination point. Nonetheless, in terms of a strict relationship between the degree of fallout and number of cancers some apparent discrepancies emerge. The detailed map on radiation fallout produced by republican specialists indicates that the population of Mahileu Oblast, after Homel', received the heaviest radiation dosage. Why then does Mahileu Oblast not appear more prominently in Table 4.3? One

Table 4.3   *Thyroid gland cancer in children in the Republic of Belarus*
(number of cases diagnosed per year)

Region	86	87	88	89	90	91	92	93	94	Total
Brest	0	0	1	1	7	5	17	24	21	76
Vitsebsk	0	0	0	0	1	3	2	0	1	7
Homel'	1	2	1	3	14	44	34	36	44	179
Hrodna	1	1	1	2	0	2	4	3	5	19
Minsk	0	1	1	1	1	1	4	3	6	18
Mahileu	0	0	0	0	2	2	1	7	4	16
City of Minsk	0	0	1	0	4	2	4	6	1	18
Belarus	2	4	5	7	29	59	66	79	82	333

Source: Demidchik, 1995.

possible answer is that iodine fallout for the most part had a significant impact on the regions close to the border, i.e. around the damaged reactor itself. This statement is also borne out by the sudden rise in incidence of TGC in the northern regions of Ukraine and in the Bryansk Oblast of Russia.[97] Demidchik himself cited the distribution of potassium iodide tablets to the population in Mahileu Oblast, based on local initiatives, immediately after the Chernobyl accident.[98] Nevertheless, the years 1993–4 saw a rise in incidence here also, with 11 of the 16 cases occurring at that time. It may thus be premature to state that there is no thyroid gland cancer among children in this oblast.

Are all the 333 cases a result of exposure to radiation? One Western radiation specialist accepts the general thesis proposed by Likhtarev and his colleagues that the number of such cases will rise to 1.4 times the normal amount. On this basis he calculates that of the 527 cases of which he was aware at the time of writing, 'about 150 were induced by radiation fallout.'[99] Another expert, Dillwyn Williams of Addenbrooke's Hospital in Cambridge, though cautious, appears to be more forthcoming in the designation of a link between the cancers and radiation. At one meeting of international experts, he points out, a straw poll was held involving 11 people, and the result was 'unanimity that the diagnosis of the tumors had been adequately substantiated and that exposure to the Chernobyl accident was definitely or probably the cause of the increase.' In his view, the probable cause of the increase, not accompanied by a concomitant rise in leukemia, was the content of radioiodine in the fallout.[100] Scientists and medical specialists who have examined the health consequences of

Chernobyl are also in agreement that increased surveillance of the health of the population is not a prime factor in the rise in incidence. Demidchik, as noted above, has been working on the field of TGC among children in the Republic of Belarus for some thirty years. When answering the question whether some cases could have existed prior to Chernobyl without detection, he responded as follows:

> Before Chernobyl, commencing in 1966, we began our medical research on thyroid cancer, and every patient with even a suspicion of this illness was tested and examined. Certainly there may have been the odd error or omission, but only in 2–3 cases at most. ... Concerning the difference between leukemia and thyroid cancer, I can say that we have accurate information about the number of cases both before and after Chernobyl.[101]

It seems safe to declare therefore that this form of cancer, thus far, has been the principal discernible consequence of the exposure of the Belarusian population to radiation. In Homel', the incidence rate had risen by 1995 to about 100 per million children per year, or 100 times the pre-Chernobyl average.

Interestingly, figures for this type of illness in the other republics affected by Chernobyl – Ukraine and Russia – are significantly lower. By the end of 1994 in Russia, for example, a total of 46 thyroid gland cancer cases had been registered, of which only seven were children. In Ukraine, according to Demidchik, 144 cases among adults and children have been registered, although because of the lack of a centralized registration system in that country, it is difficult to determine how many of these are children.[102] In Ukraine one can say that the numbers are troubling, but have not approached the situation in Belarus.

Does the development of thyroid cancer among children in Belarus constitute a crisis? In theory, this type of cancer has a very high rate of cure. Williams, for example, comments that 'in overall public health terms' the outbreaks do not yet constitute a serious problem. Presumably his reference is to the numbers involved. He acknowledges that based on the example of those exposed to nuclear testing on the Marshall Islands, such cancers could appear decades after exposure.[103] Thyroid gland cancer is surgically correctible, however. Demidchik remarked in the fall of 1994 that of 225 children (approximately an equal number of males and females) found to have such tumors, 215 were treated in Belarus and nine abroad, in addition to the seven-year-old girl who died without being treated from 'complications of pulmonary mestastases.' Among those on whom surgery was performed, 93 percent were treated for papillary carci-

noma. Past experience had revealed that the younger the sick, the better
the results of the treatment. In the past, among the age group 15–34, the
10-year survival rate in the republic was 100 percent.[104] The prognosis
therefore appears to be good. However, Demidchik has pointed out that
the cases in Belarus present different and more complex problems. In two
interviews of 1993, he elaborated on these problems in depth.[105]

He noted first of all that in the medical world there was as yet no consensus on how to treat children with thyroid cancers, and that the question
had been subject to intensive debate. The main issue is whether to extract
the entire thyroid gland, even when the tumor is small, i.e. in only half the
gland. In this case, the Belarusians usually take out the half of the gland
that is affected. In Germany and the United States, however, the entire
gland is removed.[106] The Belarusian point of view is that if the healthy
part of the gland is left within the organism, then there is a possibility of
maintaining the natural growth of the child. In other countries, doctors
used hormones to try to compensate for the loss of the gland. There were
also smaller differences concerning the use of radiation in treatment. At
prophylactic points, the Belarusians refrained from the use of radioactive
iodine if no changes in the form of cancer had been detected. It would be
applied only in the case of metastasis of the cancer. In international hospitals, radioiodine therapy is usually applied in all cases. 'We consider such
treatment,' declared Demidchik, 'to represent too much of a radiation load
on the human organism.'

If the method of treatment in Belarus is different from that abroad,
then in theory the Minsk clinic run by Demidchik must be responsible for
the vast majority of surgical operations. Demidchik maintained that
his clinic was up to such a task. He cited some three world centers for
treatment of thyroid gland cancer in Sweden, USA, and France. The
Minsk clinic had been responsible for over 500 operations by 1994, a
much larger figure than those for lung and intestinal cancers. However, it
maintained ties with centers in Japan, Switzerland, and Germany, and
links were being established with Sweden. The World Health
Organization had already signed an agreement to open an international
endrocrinological center in Minsk. But would it not be more appropriate,
Demidchik was asked, to send children abroad for treatment? Surely the
scale of the problem might soon be beyond the medical and financial
capabilities of the small republic? Demidchik's response was that those
children who were treated outside the country would require monitoring; thus repeat visits to the treatment center in question would be necessary. Such trips would entail significant costs and would impose a
burden on the children. However, the concept of treatment abroad had

already been put into practice on a small scale, partly as a result of severe shortages of medical technology in Belarus.

Thus a portion of the Belarusian children affected with TGC are being treated at a hospital in Essen, Germany. The treatment there is radioiodine therapy. The numbers, however, remain small, perhaps two or three children per month. After the treatment, the children are kept in Germany for a period of rehabilitation. There were no plans in late 1993 to send children elsewhere. The hospital in Germany was selected because the Belarusian specialists had observed that the equipment there was more modern than that in Belarus, and in fact some devices were not available domestically. In Belarus there were serious difficulties in providing children treated for TGC with the necessary medicines, sutures, adequate electrolytes, and other equipment. Demidchik asserted that the dissolution of the Soviet Union had had a major impact in this regard on all the hospitals in the republic.

Were there any dilemmas with the cancer itself? Demidchik declared that thyroid gland cancers, though curable, were particularly aggressive in form. Within two to four months, the patients suffer mestastases to the lymph lenses, or to the lungs and bones. The growth of the tumor is so rapid that the diagnosis and surgery must occur in rapid succession. Of the 215 patients who underwent operations in the period 1986–93, only 24 were at the initial stage of the illness, 142 had experienced metastasis into the lymphatic nerve centers, and nine into the lungs. Sometimes second and even third surgical operations are necessary. Looking at the results only from the period 1986–92, encompassing 162 children, in 32 cases the tumors reappeared. Recidivist cancer in the residues of thyroid tissue was noted in eight cases; and metastasis into the lymphatic system in nine cases. Repeat operations (thyroidectomies) were conducted on these patients. Fifteen children had experienced metastasis of the cancer into the lungs and were given radioiodine treatment. Demidchik and his eleven surgeons are clearly faced with a severe test of their stamina and resources. Demidchik himself has painted a portrait of a grim medical future in which thyroid gland cancers would be manifested more among teenagers and young people.

OTHER STUDIES

As noted above, the Institute of Radiation Medicine was given the main responsibility for republican inquiry into the medical effects of Chernobyl. Founded in 1986, it is led today by A.N. Stozharov, Chairman of the

Department of Radiation Medicine and Ecology at the Minsk Medical Institute. It possesses branches in Homel' and Mahileu, with its main headquarters in Minsk. It also has a branch section in the northern Vitsebsk region. The Institute has conducted clinical research, carried out radiation measurements in the territories affected by radiation fallout (calculating the radiation dosage of the population) and participated in projects conducted by other medical institutions in Belarus and the West. The latter includes projects on the state of the thyroid gland of the affected population, epidemiology of the thyroid gland, and genetic defects and malformations (again in cooperation with specialists from Germany, the United States, and Japan). Some initial conclusions to these studies were presented by Stozharov,[107] and included a rise in the general morbidity of the population which was 'probably' linked to the weakening of the immune system, though more data were needed for corroboration.[108]

The Institute is also heading an investigation into the effects of radioactive cesium on the exposed population. A study group of 5000 people was expected to increase to 10 000, with annual testing. Experiments have been conducted on rats, with the preliminary conclusion that exposure of the rodents to high levels of cesium seriously damages the capability of their bodies to repair damaged DNA. Though at an early stage, the significance of such research is evident. The effects of relatively low doses of cesium may constitute the most significant results of the Chernobyl disaster over the long term. Which groups have been most affected? According to N.A. Gres', the Director of the Endocrinological Center of the Institute, there are five major categories:

1. Children who were under the age of three at the time of the accident and living in contaminated areas;
2. Children born to women in these areas who were pregnant at the time of contamination;
3. Children whose parents took part in the campaign 'to liquidate the consequences' of the accident;
4. Children evacuated from the 30-kilometer zone around the destroyed fourth reactor unit;
5. A more recent group consisting of children living in territories with levels of 1–5 curies of cesium in the soil.

The latter group has accumulated high levels of cesium in the body and is examined with a device called a 'cesium human body counter,' and through urine tests to determine the amount of cesium in the blood. These children live mainly in Homel' and the eastern part of Brest oblasts, where the sandy peat soil is particularly retentive of radioactive particles. The

Institute has established a data base of 500 children with abnormalities and changes in the intestines and in the mucus of the stomach, which is compared to a control group living in a noncontaminated area.[109]

CONCLUSIONS

The foregoing picture of the health situation in Belarus is inevitably incomplete. There are omissions in data in various areas. Concerning the psychological effects of the disaster, the conclusions can only be tentative. The discipline itself developed only in recent years in the republic and some of the early efforts appear somewhat amateurish. If one looks at the overall picture, there can be no doubt that Belarus is a nation with a significant health crisis. Tours of hospitals and clinics on three separate occasions made several factors apparent. First there does not seem to be any overall body coordinating health information. There is no computer bank containing a register of the majority of Chernobyl victims. There are significant differences of opinion among various specialists; and there remains a wide breach between the specialists themselves and the population as a whole. As we have observed, Chernobyl affected first and foremost rural regions of the country. As a result of the disproportionate development of town and village in the Soviet period, the latter was ill-equipped to deal with an accident of such dimensions. Moreover, the enforced movement of a large population from village to town in itself may have had a significant psychological impact on those relocated.

Another evident problem noted during tours of hospitals was the general shortage of modern equipment. One Western doctor declared that the Belarusians appeared to be about three decades behind Western hospitals and some of the instruments used were obsolete in the West. He also noted that the hospitals need not have been so crowded had patients, particularly children, not been permitted such lengthy periods of recuperation following surgical procedure.[110] The Chief Physician of the city of Minsk pointed out some of the problems with vaccines in the city, most of which come from Russia and do not conform to international standards. The Russian vaccine currently used for immunization of children against diphtheria has resulted in periods of quite severe illness and has not been certified internationally. In addition, there are as yet no tuberculosis vaccines for newborns, despite the prevalence of that disease in the former Soviet Union.[111]

Some of these dilemmas can be seen within the context of the transition from Soviet rule to independence, and in the case of Belarus, the continued failures of self-assertion as a national state that can clearly be demarcated from Russia. Other weaknesses of the state medical system include the lack of knowledge about medical rehabilitation, Sudden Infant Death Syndrome (SIDS), the shortage of family doctors – this concept was virtually unknown under the Soviet system – and a serious lack of vital equipment, including respirators, anesthetics, vaccines, and drugs generally. All these deficiencies apply to the republican health system generally. Observations in Minsk may not be typical: one might assume that the capital city would receive more attention than the average location, though perhaps no more today than an irradiated regional center under international medical surveillance such as Homel'.

These material factors need to be considered alongside a more undefinable malaise: a general apathy and lack of confidence among the population toward existing state organs; a feeling or perception that life is not going to improve and that children in particular face an uncertain future in the state of their health. One can posit that such sentiments would be passed on to the children, and that the latter would come quickly to the realization of the woeful state of affairs in the republic after visits abroad.[112] As Grushevoy has noted, there is at the same time an almost pathetic reliance on the aid of the state that has become part of the national psyche. That state, in turn, has concerns other than Chernobyl.[113] The economic picture at the end of 1994 can accurately be described as disastrous: GDP and industrial production had both dropped by 22 percent compared to the output figures for the previous year; the production of consumer goods fell by 17 percent, while consumer prices soared by 25 times. The new government that took office in the summer of 1994 found itself with a debt of some one billion US$, almost half of which was made up of gas debts to Russia. It was estimated, in addition, that almost 14 percent of budgetary expenses for 1995 would be devoted to 'liquidating the consequences of the Chernobyl disaster.'[114] Consequently, the victims of Chernobyl may not feel forgotten, but given other factors they are dependent to an extraordinary degree on international aid, from charitable associations, Christian groups and others. Ironically, given twentieth-century history, the major participant in such aid is Germany. In private conversation some individual, mainly elderly, Belarusians expressed both confusion and disgust at this turn of affairs – that they had become reliant on the former occupants of their land for survival.

Can one agree with Dillwyn Williams that the current health problems deriving from Chernobyl do not as yet constitute a crisis? The answer is that those problems relating to Chernobyl cannot realistically be separated from the health situation in the republic generally. The problems in health care have belatedly become the focus of official attention, that is they have become so paramount to the public that the government and parliament are obliged to act, or at least to be perceived to be paying attention to such matters. Hospitals possess outdated equipment; hospital beds are full to capacity; diseases such as diabetes are striking children at a much earlier age than hitherto; illnesses that are almost unknown in Western Europe, such as tuberculosis and diphtheria have resurfaced in the republic. We have outlined above the shortage of drugs and suitable vaccines for children. In January 1995, Kazakov was removed from office and replaced with I. Drabysheuskaya. Yet, one source pointed out, Kazakov was not entirely to blame for the problems that have arisen. He was not working for personal ends, but for a program devised by the Supreme Soviet, a body not prepared to tolerate new ideas or to adopt programs based on world experience in health care. Discussion among the Association of Doctors of the Republic of Belarus focused on a medical care system outside state control.[115]

The Chernobyl factor has played a major role in this crisis. It has certainly opened up the republic to international scrutiny of its health care system, while the presence at international conferences of leading doctors such as Demidchik and Ivanov has also raised awareness of local inadequacies. Moreover, whether or not one considers the health effects of the accident themselves to constitute a crisis, there can be no doubt that they have produced an enormous burden on the existing health system. Many of the illnesses that pervade the contaminated zones may be psychosomatic, others may be unrelated to increased radiation levels in the soil, or in the air in the summer of 1986, but this is not to say that they do not exist. Moreover, there has been a rise in diseases generally, including leukemia. Finally, the thyroid cancer outbreak itself, especially among children, is cause for serious concern, particularly since the numbers continued to rise in 1994. One can declare that the forecast of Il'in and others of a 1.4 rise in numbers has proven to be flawed. In the affected regions, levels are 80–100 times the norm and continuing to increase. Though not usually fatal, thyroid gland cancer has been seen to be a highly aggressive disease. Those affected will require monitoring and treatment for the rest of their lives.

The first ten years after Chernobyl, then, have posed considerable problems for the government of Belarus. Statistically, the increase in morbidity among the general population is quite dramatic. There is no justification for dismissing such an increase as the result either of outbreaks of radiophobia or as a result of improved methods of detection of illnesses. (Both these factors assuredly exist, nonetheless.) The number of cancers of all types has risen, though only thyroid gland cancers among children have increased to levels significantly above the world average. Thyroid gland cancer at such levels was unexpected. There were no predictions of this occurrence by the medical and scientific community in 1986. These disturbing medical problems have been compounded by what can genuinely be termed a crisis of health care. Though international aid has been significant and essential, it cannot compensate for the absence of a viable and modern domestic health care system. At the republican level, it was evident during several visits to the republic that coordination of health care was minimal. Indeed, most of the administrative and practical supervision over hospitals and clinics was left to the city authorities.

Finally, it would be a mistake to underestimate the mood of the public on the question of the health effects of Chernobyl. The general air of hopelessness that pervaded the affected has become clearly discernible to the outside observer. Ultimately, the public lacks faith in the present health care system. The Lukashenka administration and the Belarusian Supreme Soviet are aware of the problem. Whether it can be resolved at the state level is another matter. Ideally, the initiatives taken by the more viable charitable organizations might develop into a privatized health care system that can take over some of the burden of care for Chernobyl and other health factors among the population. In early 1995, however, such a process was neither taking place nor anticipated in the near future.

5 Perestroika and Independence

The number of books written about the last years of the Soviet Union is probably less in the post-Soviet period than in the period 1988–91. For a time during the late 1980s, a spate of books focused on Gorbachev and his reforms; particularly the Soviet economy and the revelations of glasnost. Perhaps publishers have become more wary. The sudden end of the Soviet Union left many of them with a glut of unsold works that concentrated for the most part on prognostications on what the former leader might do and where the USSR might be in five or ten years' time. Thus it may appear as though I were taking a step backward in making reference once again to the period of perestroika – defined somewhat inaccurately as embracing the years 1985–91.

The intention, however, is not to discuss perestroika *per se*, but rather to formulate an argument about the specific characteristics of the development of Belarus to conclude this monograph. In discussing the Gorbachev era, most works see several turning points: periods after which the Gorbachev regime took – to use a Maoist phrase – 'great leaps forward.' The first of these appears indubitably to be the Chernobyl disaster, which occurred only thirteen months after Gorbachev became the General Secretary of the CC CPSU. After this event, it can be argued, one saw not merely perestroika imposed from above, but within two to three years a grassroots movement for perestroika from below. It could be argued that after this occurrence, events took their own course. Gorbachev was the captain of a ship that was soon out of control.

If Chernobyl was the catalyst for the development of glasnost, there were other factors that enhanced its progress: the development of public opinion manifested itself in an outpouring of literature; the ability to study the past led to the posthumous rehabilitation of most of the victims of the Stalin purges; the ending of the war in Afghanistan; the 19th Party Conference in the summer of 1988, and Gorbachev's decision to call elections for a Congress of Deputies later that year (with elections scheduled for the following March). These events combined to create movements oriented away from the Moscow center that had initiated them. They were no longer led by the Communist Party of the Soviet Union. Though the party and KGB remained powerful institutions, it was possible for a grassroots movement to survive even with their opposition. Finally, perhaps the USSR's decision not to oppose the removal of Communist regimes in Eastern Europe was the critical factor: it paved the way for popular movements to act in similar fashion within Soviet borders, albeit to a much more limited extent.

In retrospect, it is perhaps surprising that after the Baltic states, the Slavic republics should take the lead in this process. In the Brezhnev years, the republic of Belarus had been increasingly compliant, stifling any local initiatives and awaiting directions from Moscow. Chernobyl and its ramifications came as a profound psychological shock to the republic. Its leaders were incapable of formulating an appropriate response. After 2 May 1986 – six days after the accident – matters were transferred to a Government Commission run by a deputy chairman of the USSR Council of Ministers, two Politburo officials – Nikolay Ryzhkov and Evgeniy Ligachev – who supervised the initial evacuation of a 30-kilometer zone, the USSR State Committee for Hydrometeorology and Environmental Control, the USSR Ministry of Medium Machine Building, and the Moscow-based Academy of Sciences. In short, the operation was in the hands of the centralized state structure rather than republican organizations. By 1989 in Minsk, there was widespread and openly expressed disaffection with the lack of initiative at the republican level; the centralization of control over republican industries; and also toward organizations such as the International Atomic Energy Agency (IAEA), which was permitted an important advisory role after Chernobyl but appeared to many observers to adhere too strongly to a pro-nuclear power stance.[1]

Following the example of the Baltic states, a Popular Front was formed in both Belarus and Ukraine that served as an alternative voice to that of the Communist Party hierarchy. There the similarity between the two republics ends, however: in Ukraine the Rukh developed into a vocal and progressive movement that was to take Ukraine to sovereignty by 1990 and to play a significant role in the changes that followed. That it could achieve such results was due in part to its incorporation of a significant group of Communist Party members, particularly in the parliament. In Belarus, by contrast, the Communists and Popular Front could find no common ground. Though sovereignty was formally proclaimed there also in the summer of 1990, the populists were unable to gain a significant foothold in political life and the Communist bureaucracy withstood the early shock and recovered its control over society. As described by one Minsk observer:

This newcomer to Belarusian politics [the Belarusian Popular Front] did not enjoy as much popularity as similar fronts in the Baltics. Such feeble support, politicians think, may be due to the fact that nationalist sentiments in Belarus are considerably weaker.[2]

The observer can trace the end of Communist rule and the development of democracy in Ukraine directly to the response to the Chernobyl disaster. The population as a whole could see clearly the follies of a centralized system, of outside control over natural resources and industrial enterprises, such as nuclear power stations, chemical plants and metallurgical works. If Chernobyl did not inspire nationalism directly, it certainly provoked a patriotic response over the next few years. From 1990 to the summer of 1991, the infrastructure of a national state took shape as the Ukrainian Supreme Soviet adopted several radical laws. At Kyiv City Hall on the Kreshchatyk in Kyiv, the blue and yellow flag replaced the Communist emblem. A former party secretary for ideology, Leonid Kravchuk, became – seemingly overnight – a forthright and determined patriot prepared to defend to the bitter end what he perceived as Ukraine's national interests. These facts are well known. Yet why did the small republic to the north not respond in a similar manner? Why did the patriotic movement encapsulated by the Belarusian Popular Front (BPF) fail manifestly to capture the attention of the public to the same degree? Was it merely because, as lamented by Zyanon Paz'nyak, Belarus lacked a politician of the stature of Kravchuk? Or that Belarus lacks the natural resources of Ukraine? Or were there deeper, underlying reasons?

This study has emphasized a theoretical premise: namely that the crisis of Chernobyl in Belarus has been exacerbated, possibly even partially created, by the lack of national state development in the nineteenth and early twentieth centuries, and particularly in the Soviet period when an alternative form of nationalism was espoused (see below), and further that the present state lacks some of the basic prerequisites of an independent nation: a developed national consciousness; a broad urbanized intellectual national elite that has bases throughout the country; and a population that freely and fluently uses its national language. While a critical observer might choose to chastize Paz'nyak as a potential national leader on some points – for example, what is widely perceived as his Russophobia and a lack of tolerance for Russophones generally; an occasional tendency toward hyperbole in his speeches and writings – the problem is not his, but rather one firmly rooted in modern Belarusian history. At the same time, the Soviet period should be seen as the key one. Again, the example of Ukraine provides an alternative mode of development.

In the early twentieth century, Ukraine, like Belarus, was a predominantly rural society, both in the Austrian and Russian-ruled areas. The Soviet period could with justice be perceived as an unmitigated disaster for Ukraine: it resulted in three serious famines (1921–3, 1932–3, and 1946–7), mass purges, and in the 1930s the systematic elimination of a national elite. It saw the period of German occupation, and several years of bitter conflict in the western regions between nationalist guerrillas and Soviet forces in the postwar era. At the same time, the Soviet period did much to unify Ukraine politically. It saw the annexation of Western Ukraine to the Ukrainian SSR, and the postwar incorporation of Transcarpathian Ukraine from Czechoslovakia after the war. Soviet sources constantly used the phrase 'reunification' to describe these events, which may also be perceived as machiavellian territory-grabbing on the part of the Stalinist regime. Nonetheless, both territorial changes and Soviet economic policies served to enhance rather than reduce national consciousness on the part of Ukrainians.

Ukraine possessed some significant advantages over Belarus. Its population was approximately five times larger. It possessed a more significant agrarian and industrial base. In Western Ukraine even the peasantry had a tradition of self-help and cooperative societies. The Ukrainian language, which had thrived only in the villages of the western regions, was brought to the industrial centers by the migrating peasants involved in the industrialization campaigns of the 1930s.[3] Stalin's policies, despite the brutality with which they were executed, served to develop an urban Ukrainian elite. The significance of the city of L'viv and the L'viv region generally in this process should not be underestimated. Even under Polish rule in the interwar period, it saw the development of an at times fanatical integral nationalism. On the other hand, the role of the city of Kyiv and the former Ukrainian territories of the Russian Empire is all too often underrated as a participant in this process of nation building. Russians in Ukraine are Russian Ukrainians rather than Ukrainian Russians, that is they are part of a Ukrainian state and for the most part divorced from their ancestral motherland, Russia. Whatever the limitations of national state development in twentieth century Ukraine, progress had been made by the Gorbachev period.

In Belarus, it is posited, the twentieth century has served to weaken severely the last vestiges of a national form of state development. The observer could argue that such nationalist sentiment never existed in the republic. This is not the case. As illustrated by the short-lived cultural development of the 1920s, the Belarusians could make rapid strides when given official encouragement and investment in native-language usage in schools and official meetings. The Stalin period in particular – though both the Khrushchev and Brezhnev periods played an important role – saw the destruction of the potential national state, partly by design and, one is charitably inclined to assert, partly by accident. There was no predetermined genocide of the Belarusians as a nation, though some writers have argued that the destruction of the Belarusian intellectual elite in the 1930s was tantamount to such a policy; through its policies, let us say, the Soviet state eliminated a fledgling movement. How could the USSR achieve the elimination of national development in its third Slavic republic, given that the state could hardly ignore or remove some eight million citizens of Belarusian ancestry?

A simple answer presents itself. Soviet nationalism superseded and served to eliminate Belarusian nationalism. What is Soviet nationalism? Is it different from or related to Russian nationalism? Surely even in Russia, the heartland of the Soviet system, it is possible to elaborate a viewpoint that over the course of 70 years the USSR not only failed to eradicate Russian nationalism but even began to redevelop it as a response to Hitler's invasion of the USSR and beyond? The concept of the New Soviet Man could be perceived as a mythical creation and has been the subject of ridicule. Writing in 1985, Paul B. Henze noted that:

> The Russians at most levels in Soviet society are increasingly aware of the pain and cost of operating a multi-national empire. Like all imperial peoples, they find that benevolent treatment of subject nations does not produce gratitude. Raising these peoples to a higher level of economic development, opening up new social and intellectual horizons for them, does not necessarily make mutual relationships easier. Some Russians react with scorn and resentment; some with open hostility; some with patience and dedication to principle; many with indifference. Russians have not developed a successful formula for turning non-Russians into Russians. Officially they have to keep denying that they are trying to do so at all. New Soviet Man has proved elusive.[4]

Yet in Belarus there is evidence to suggest that the concept of New Soviet Man took on real meaning for several reasons. First, the national intelligentsia was small enough to have been almost totally eradicated in 1937–41. Second, the worst excesses of the Stalin years, including the purges and the mass executions as at Kurapaty, were kept secret from the bulk of the population, and even today are either not well known or officially denied. Third, the occupation of Belarus by the Germans occurred at the very beginning of the Great Patriotic War and was the longest such occupation and perhaps the most brutal era in twentieth-century Belarusian history. It left its mark on all the population. The arrival of the Soviet army, combined with local efforts of partisans who took their directions from the NKVD in Moscow, was perceived by many sectors of society as a genuine liberation. The Soviet Union had 'saved' Belarus, which had lost 25 percent of its population in the war. Michael Urban and Jan Zaprudnik observe that:

> The Second World War represented a pivotal experience in the making of the Belarusian nation ... This wartime experience would prove seminal in shaping postwar politics in the BSSR. On the one hand, the bulk of the former partisan leadership embarked on political careers in the postwar period ... On the other hand, the partisans were able to promote at least a limited version of Belarusian identity within the BSSR. Drawing on their experiences in the resistance, they portrayed the liberation, indeed, the salvation, of Belarus as the result of the heroic *national* resistance of the Belarusian people within the larger framework of the tremendous sacrifice and achievements of the *Soviet* people.[5]

I would stress the idea of national Soviet existence rather than a version of a Belarusian national identity in the BSSR. The concept of the heroic party combined myth and reality. Resistance to Soviet rule and the extent of pro-German feeling in the republic was considerably greater than officially stated.[6] Nevertheless, pride in the Soviet motherland and a feeling of close affinity with Russia persisted both during and after the 'Great Patriotic War.'

THE BELARUSIAN POPULAR FRONT

We have argued in Chapter 1 that both Soviet economic and demographic policies served to impede state development in Belarus. Though there were some brief moments of optimism, the republic in 1985 had lost virtually all semblance of an emergent national state. Neither in its institutions nor in its expressed political sentiment could a declaration of independence be justified. The Belarusian Popular Front (BPF) remained a small and isolated force that was attacked, at times with viciousness, by the regime. It was obliged to hold its founding congress in Vilnius, Lithuania on 24–25 June 1989, while its members were denigrated at every opportunity and equated with German National Socialists in some of the most farcical yet hostile propaganda devised by the authorities. According to one hostile account, the BPF had no program, and its 400–500 delegates were ignored by the population of Vilnius, which went about its business while the Front Congress took place in the Republican House of Trade Unions. Despite the fact that the notion to form a Front had originated from cultural groups, the writer declared, none of the leading republican creative unions attended the Congress; and there was also the notable absence of leading writers and artists. Ales' Adamovich, for example, had sent only a brief telegram; Vasil Bykau was in Spain; and other Popular Front members (from Lithuania and Estonia) were thin on the ground. The impression therefore is of a small and clandestine meeting with little popular support.[7]

In another account of the founding congress of the BPF, it was noted that on 24 May, one month before the scheduled convocation, 16 BPF members, including four parliamentary deputies, asked the Presidium of the Supreme Soviet for permission 'to cooperate in the holding of the Congress.' No newspapers in Belarus reported this request until 11 June, and at that time the names of the deputies were not mentioned. Nevertheless, the BPF was publicly admonished for its alleged secrecy by the Deputy Minister of Justice of the BSSR, V. Lovchy, and the fledgling organization was accused of replacing democratic ideals with extremism. The authorities then tried to postpone the Congress. Stanislau Shushkevich, at that time the Pro-Rector of the Belarusian State University in Minsk, declared that the plan was not to ban the Congress formally, but to completely disrupt it with the use of red tape.[8]

Even before the founding Congress took place, the BPF came under attack from government organs. The movement was associated with both the founding of the Belarusian People's Republic of 15 March 1918 and a reported Second All-Belarusian Congress of June 1944 under German occupation, which was directed against Bolshevism. Both events were equated by critics with the growth of fascism, a subject guaranteed to inflame passions in the republic. The party branch of the 'central' region of Minsk saw fit to issue a special edition of its newspaper devoted entirely to critique of the BPF and its platform. An incendiary headline declared 'No! No to the enemies of the people! No to provocations! No to nationalists! Yes – to the Soviet Socialist Motherland!'[9] Within the Belarusian capital, the Communist authorities missed no opportunity to assail the populist movement at every opportunity.[10]

There were also attempts to prevent the organization of a Minsk protest by the BPF against secrecy over Chernobyl that would include among the demonstrators residents from the contaminated areas – specifically

Naroulya and Khoiniki districts of Homel' Oblast, and irradiated areas of Mahileu Oblast. These residents intended to travel to Minsk to complain that they had not been included among the areas of concern, despite the fact that some of the areas had registered levels of cesium in the soil of up to 60 curies/km^2. First, the authorities announced a *subbotnik* (working Saturday on a voluntary basis, a feature of Soviet rule) for that day, 30 September 1989. The State Auto Inspectorate (GAI) was ordered to refuse entry into the city of 'suspicious' buses from Naroulya and Khoiniki, and other contaminated areas. The demonstration, nonetheless, took place. The authorities then tried to prevent its gathering physically until the crowd swelled to about 30 000. Once assembled, they were addressed only by Evgeniy Velikhov, then Vice-President of the USSR Academy of Sciences, and by their own appointed speakers, i.e. no one from the Belarusian government saw fit to speak to them about their concerns. In the words of Ales' Adamovich, the 'ecological Kurapaty' awaited serious investigation.[11] The incident reflects the irrevocably hostile attitude of the authorities to any action that was planned or supported by the BPF. A year earlier, on 30 October 1988, another demonstration in Minsk devoted to the memory of those who died during the Stalinist repressions (a familiar event in many republics at this time) had been brutally dispersed by the militia.[12] In Minsk, no quarter was to be given to any non-Communist group.

Moreover, the activities of the BPF were largely confined to Minsk and to parts of the Hrodna and Brest regions. In none of these three areas could it be said to be clearly the dominant force. Ultimately the majority would prevail over the vocal minority. Minsk was the official capital and cultural center, but as we have demonstrated it was in essence a Russian-speaking city and the center of the Belarusian Communist Party organization. Like other Popular Fronts, the Belarusian version was supported by the Belarusian diaspora in the West, which gave added impetus to its statements and congresses, but that diaspora was not numerous enough to have a significant impact on world opinion or on local circumstances in Belarus itself. The authorities in fact had a virtually free hand to impede the BPF at every opportunity. In the parliamentary elections of 1989, it was not permitted to register officially,[13] but it managed to elect some 22 deputies who sat in the Supreme Soviet as independents. Thus it composed less than 10 percent of total deputies. The BPF took an ambiguous stand toward parliamentary speaker Stanislau Shushkevich, a man who supported democratic processes in Belarus, but lacked a power base to see such processes to fruition.[14] As a result, he was (perhaps erroneously) perceived by BPF supporters as a weak leader and one unable to assist significantly in the promotion of democracy in the republic.

The BPF itself offered an historical analysis of the movement in the republic at its Third Convention in 1993, which provides an instructive perspective. It described the political scene in Belarus as a 'snake-ridden field,' that could only be crossed without faltering or looking back. On 19 October 1988, an organizational committee had been formed. Eleven days later a bloody confrontation had been prepared in Minsk by the Communist authorities, but the BPF had survived. In February 1989, the first victory had been achieved over the 'Communist partocracy' with an overcrowded meeting of the BPF at the Dynamo Stadium in Minsk. Further progress had been made with the founding congress in Vilnius in July, and two international gatherings in September 1989 with focus on Chernobyl. Finally in February 1990, the early stage of the movement had culminated, when over 100 000 gathered in Minsk for the preelection assembly of the BPF. Subsequently, during the elections, the nomenklatura had staged a recovery by exploiting its control over the media and the juridical establishment. An outpouring of anti-BPF propaganda and falsification of truth had ensued, and had served to weaken significantly the number of electoral votes obtained by the BPF.[15]

Between late 1990 and November 1992, the opposition nonetheless made significant progress, commencing with a large-scale anti-Communist rally in Minsk on 7 November 1990. On 24–25 March 1991 the BPF held its Second Congress on the theme of the revival of national freedom and independence. On 19 July it was officially registered by the Ministry of Justice as a public political movement.[16] On 19 August, the BPF issued an appeal to resist the Emergency Committee that had been established in Moscow, and arranged a picket protest in the center of Minsk. Opposition deputies also initiated the critical emergency session of parliament that led to the declaration of independence, and the cessation of all activity of the Communist Party of the Soviet Union (CPSU) and the Communist Party of Belarus (CPB). The analysis then centred on the campaign to force a new parliamentary election through the collection of signatures for a referendum. By March 1992, over 442 000 signatures had been gathered, but the government refused to acknowledge the validity of some of the signatures, and by November 1992, had resolved not to hold a referendum under any circumstances. According to the BPF leadership, this decision was a signal that the Communists had returned to power and ended hopes that the country might embark upon a path of progressive economic reforms. The government, in the view of the BPF, had suppressed discussion on the question of neutrality, sovereignty, and the native language, and had moved toward the restoration of the Soviet system in the republic.[17]

How had the nomenklatura managed to reassert its authority? According to the BPF, there were two reasons. First, the public were fearful of the disintegration of the state infrastructure, of economic bankruptcy, and clung to the belief that a collective security treaty represented the best hope to prevent civil strife or political crisis. Second, psychologically, the populace had been firmly subjected to the state and Soviet rule. A large proportion of the workforce continued to work for the state more than three years after independence. Moreover, the apathy of the population toward its own nation, property rights, and personal liberty served to restore the confidence of the nomenklatura. Thus a change had occurred, but it was not a transition from a Communist state to a democracy. Rather Belarus had undergone a metamorphosis only from a partocracy to an administrative economy dominated by bureaucrats and controlled by black marketeers. Insofar as privatization had taken place, it benefitted only those already in power. In some respects, according to this viewpoint, the establishment structure had become more sinister, since it now encompassed elements that could be described as *mafiosi* or organized criminals.[18]

By 1994, the rift between the BPF and the parliamentary speaker was complete and in the presidential elections, Shushkevich, who had been ousted on the evidently unjustified accusation of corruption in January 1994, stood as an independent candidate against, *inter alia*, Paz'nyak of the BPF. Both were defeated in the first round of voting, and together they received only some 23 percent of the total vote. If one assumes that some of those voters who supported Shushkevich might have supported a united BPF campaign, then perhaps 15–20 percent of the electorate backed the opposition generally in support of a program geared to the development of a fundamentally Belarusian national state. The remainder of the population supported either a hardline 'old-school' Communist or (the vast majority) a young populist proto-Communist who demanded close ties with Russia and expressed his regret for the dissolution of the USSR three years earlier. As Gennadiy Grushevoy commented, the electorate was faced with a question of electing the lesser of two evils.[19]

We are not suggesting here that the current regime has no interest in specifically Belarusian concerns or in maintaining some form of theoretical independence. Indeed that would be highly unlikely for the simple reason that Belarus is the source of power for these politicians, particularly Lukashenka. To become completely subservient to Russia would reduce significantly this power base and render their roles somewhat meaningless. Rather one can say that a majority do not operate exclusively in the service of Belarusian state interests, but rather toward a policy oriented predominantly toward Russia and some form of Union with the huge neighbor state. It is fair to say that a majority of Belarusian residents support at the least an economic union; and even the policy advocated by former prime minister Vyachaslau Kebich of a military-security union with Russia would possibly have been acceptable given a national referendum. In this sense, Belarus might be described once again as a state that has a death wish. Many of its residents would be prepared to sacrifice independence if they could be assured of an improvement in their economic well-being.

A disaster of the magnitude of Chernobyl called for not merely national unity, but a national will to overcome its consequences. Ukrainian activists saw the accident as one reason why the control of Moscow over economic life should be ended. For the Belarusian state leaders (the powerless Shushkevich being an exception), the attitude was different: without the aid of Moscow, how could an event of such enormity be overcome?[20] The result was that Chernobyl had a far more adverse effect upon Belarus than elsewhere. For over three years the population remained largely ignorant of the significance of the disaster outside the Chernobyl region itself. When Chernobyl was compared in importance to the Great Patriotic War, the comparison served to promote the view that the Soviet Union would be responsible for overcoming the fallout, rather than the individual republics. In the period 1986–9, only the BPF attempted to alert the population to the dangers posed by the accident. Consequently, the Soviet leadership of Belarus, oriented primarily if not exclusively to all-Union policies, was unable or unwilling to offer a national response to a national emergency. There could be no national response without a concomitant commitment to a national statehood or, at the least, to full economic sovereignty.

Nonetheless, in August 1991, when the military junta temporarily took over the leadership of the Soviet Union in an attempted putsch, did the Belarusians finally take action to divorce themselves from the discredited Soviet state? The republic was by this time fully aware of the import of Chernobyl and the impact of official secrecy. It had been made cognizant also of some of the worst excesses of the Stalinist years. The answer is that 22 members of the opposition issued a declaration condemning the coup on 19 August (the first day) and demanded an emergency session of the parliament. While such action was significant and exceptionally brave, it indicates the minute size of the democratic movement within the highest state body. These same 22 members also put on the agenda of the emergency session the questions of abrogation of the 1922 Union Treaty and the declaration of independence in Belarus. It was the Communists, however, who recommended that the independence declaration be accepted; this step was considered as a matter of personal survival, given the collapse of the putsch and subsequent banning of the Communist Party.[21]

Four months later there occurred one of the most bizarre events of the tumultuous 1991 year, namely the formation of the Commonwealth of Independent States (CIS) in Brest Oblast, on the initiative of Boris Yeltsin, but with the active participation of Shushkevich, and the newly elected Ukrainian president, Leonid Kravchuk. The three Slavic republics made common cause, ostensibly against the concept of a revived union in which Soviet President Mikhail Gorbachev could still conceivably play a role of symbolic significance. Why would Belarus and Ukraine make such a commitment? In Ukraine's case, the answer may be a simple one: there was no authentic commitment by Kravchuk to anything other than the loosest federative ties. Ukraine was never fully committed to the CIS. For Belarus, and for Minsk as the 'capital' of the CIS, the accord was perhaps more sigificant. For Shushkevich, it may have been an attempt to bolster his support at home for an independent state that was no longer linked to the Soviet Union. For the majority of Communists in the Supreme Soviet, on the other hand, it may have signified that links to Russia were far from broken. It could be argued that the independence of the state was at once compromised by the formation of the CIS.

Moreover, the independence of Belarus can be considered an event that was largely determined by external circumstances. It originated directly with the events in Moscow and would not have occurred without them. Though economic sovereignty had occurred the previous summer, this decree was largely a paper one. The vast majority of decisions on the Belarusian economy were made in Moscow. More important, Belarus was the most significant and secretive Soviet military base and the most militarized of the Soviet republics. (Indeed the declaration of independence stranded thousands of Soviet troops on Belarusian territory.) Belarus found itself with quasi-Soviet leadership in the post-Soviet period. The mentality of the leaders did not change simply with a change of political status. Kebich, Myachaslau Hryb, and others could have taken steps to develop local initiatives in dealing with Chernobyl. They could have supported actively republican charitable funds to alleviate social problems. In fact they chose to do the opposite: to attempt to stifle regional initiatives on the pretext that they were politically motivated,[22] and only in the most dire circumstances did they establish the controversial State Chernobyl Committee that soon became a target of Lukashenka's anti-corruption committee.

The goal of this monograph therefore has been not merely to highlight the health impact of Chernobyl upon a republic, though this possibly constitutes the most significant part of the work. It is also to demonstrate that the problems engendered by that accident cannot be seen in isolation from the way in which the state was developed economically and demographically. The question of the relative lack of national consciousness generally in Belarus must also be taken into consideration. Belarusians, as a whole, neither acted nor responded as a nation. Those that were capable of such a response were not in positions of authority. An emergent nation might have responded to a disaster of such magnitude by throwing off the former authorities and developing a new regime oriented away from Moscow. The Republic of Belarus was not in this situation. As a nation thoroughly integrated into the Soviet order, as a compliant republic of the Soviet Union, its leaders awaited orders, as in the past. The measures adopted hastily and belatedly since 1991 were far from comprehensive. In many respects they echoed measures taken in Ukraine.

Nonetheless the authority of the state bureaucracy thus far has been sufficient to quash local initiatives also. It is a truism that virtually every significant nongovernment attempt to deal with the effects of Chernobyl has emanated from organizations or individuals who can be equated with the political opposition in parliament, or with the BPF. These people contain some patriots and some very talented and inspired individuals, but there is within them no one individual, or no one group that has significant political authority or stature. By 1995, as the world approached the tenth anniversary of Chernobyl, the situation within Belarus appeared to have deteriorated with time: the economic and social crisis had solidified; the new president and his parliament were engaged in a power struggle; and the proportion of the budget that was to be devoted specifically to Chernobyl had been reduced to 10 percent[23] – still a significant sum, but one that indicates that the authorities have resolved to reduce their commitment to those affected by the accident at a time when medical and social questions are at their most acute level since 1986. If Chernobyl can be termed a tragedy, then part of that tragedy is the failure of existing state organs to offer an adequate and lasting response to that event.

THE NUCLEAR POWER QUESTION

A macabre postscript to the questions arising from Chernobyl has been the debate within the leadership of Belarus on the question of constructing nuclear power stations in the republic. The debate has reflected both the influence of Russia – which announced its own new program for long-term nuclear power station construction in December 1992 – and the extraordinary insensitivity of the state leaders to the problems brought by the nuclear accident. That Belarus, a republic without any nuclear power stations, could embark upon a program to build such plants so soon after the misfortunes caused by a previous nuclear accident, albeit from a station just over the border in Ukraine, astonished many observers. As the most significant figure in Belarusian political life in 1992, parliamentary speaker Stanislau Shushkevich, a physicist by training, gave his assent to the plans. In his view, Belarusian stations could be constructed using foreign technology. Already by April 1992, he had held talks on the subject with nuclear experts from France, Canada, and Russia.[24]

What arguments have been used to support such a program? Two proponents of nuclear energy are G. Koren' and A. Yaroshevich. Koren' is head of a production-technical section of the organization 'Belarus Energy,' while Yaroshevich is head of laboratories of the Institute of Problems of Energetics with the Belarusian Academy of Sciences. They outlined their arguments in an issue of the journal *Inzhernerskaya gazeta*.[25] First, they noted, the basic source of energy in the republic consists of organic types of fuel. These account for only 12 percent of the needs of fuel-energy. The remaining 88 percent must be imported. Thus Belarus suffers from an acute fuel deficit and is almost totally reliant on imports. In addition, the organic fuel production has contributed to the deteriorating environmental situation in the republic as a result of waste products. The need for alternative energy sources, in their view, is clear.

What are these alternative sources? In their view, extensive research at scientific institutes has demonstrated that the use of wind-based energy, solar energy, energy from rivers and bioenergetics could not significantly reduce the dependence on energy imports, The development of energy saving could reduce the necessary tempo for energy production, but it could not resolve the principal problem: the deficit of energy resources in Belarus. Another author has examined these alternatives in more depth. I. Edchik, a senior scientific worker with the Institute of Problems of Energetics, affiliated with the Belarusian Academy of Sciences, has examined and dismissed each alternative in turn. Wind energy is not feasible because the average wind speed in the republic (about 14.4 kilometers per hour) is insufficient, and would permit at best the use of only a small portion of wind energy (some 1.5–2.5 percent of all energy needs). Solar energy also seems impractical. Its lack of general intensity, the variable weather, the difficulties in providing land for solar energy all present great difficulties for the harnessing of this form of energy. Bioenergetics – the use of organic deposits – could provide at best the equivalent of 2–3 million tons of conventional fuel (5–7 percent of total needs). Finally the energy of small rivers and waterfalls produces only the equivalent of 0.1 million tons of conventional fuel.[26] On all these counts, Edchik's study supports that of Koren' and Yaroshevich.

Returning to the latter, they point out that the concept of nuclear power stations in Belarus is not new. Indeed it was already under way during the last Soviet five-year plan. After Chernobyl, however, the program was curtailed and work on the construction of a nuclear power and heating station near Minsk (Minsk ATETs) was halted in 1988, while plans for a Belarusian nuclear power station were abandoned. The authors do not ignore the question of Chernobyl and raise the question of whether in the conditions in which the republic finds itself, as the most contaminated as a result of the Chernobyl disaster, public opinion would countenance the building of nuclear power stations. They must study the question 'with the minimum of emotions', the authors respond to their own rhetorical question. Chernobyl led to the slowdown of nuclear energy development worldwide. Various measures were introduced to improve the reliability and safety of stations. This procedure took time and material resources. Yet the majority of countries did not fundamentally change their long-term energy programs. Rather they maximized the safety of existing nuclear stations before continuing to construct more.[27]

In nuclear energy production, the authors stressed, one did not produce the release of harmful substances into the atmosphere such as carbon dioxide, sulfur dioxide or nitrogen oxide, and there was no burning of oxygen as in organic-fuelled stations. Under normal circumstances, the release of radioactive substances was substantially lower than in the case of thermal power stations, especially those based on coal. Thus it would be 'incorrect' to reject the nuclear power alternative, Koren' and Yaroshevich assert. Yet how could one guarantee that the already damaged republic would not be subject to future accidents, either on the Chernobyl scale or more minor in nature? The answer lay in the selection of the type of reactor for the station. The station should not have Chernobyl-type reactors (Soviet RBMKs) or VVER (water-pressurized) reactors used elsewhere in the former Soviet Union. In Russia and elsewhere, a new generation of reactors was under construction. After the year 2000 another type of reactor with its own 'internal safety system' would be ready for use. In Belarus, the selection should take the form of a competition, based on the views of local and international experts. Russian nuclear plants and those further abroad would have to be investigated. All Belarusian reactors must have two protective covers to alleviate the potential damage of a release of radioactive products.

Though nuclear fuel would still have to be purchased from Russia, the authors maintain that the costs would still be much lower than for organic fuel imports. In fact the cost of such fuel has declined as a result of the reduction of military development and the conversion of weapons enterprises into civilian factories. In conclusion, the authors declared, there is no alternative to the development of nuclear power for Belarus, whatever the emotional implications of such a move. Their optimism is shared by Aleksandr Mikhalevich, Director of the Institute for the Problems of Energetics at the Belarusian Academy of Sciences. There was, in his view, no time for despair after Chernobyl. Scientists began to research new types of reactors and today the danger of a Chernobyl-type accident was 1000 times less than in 1986. The world had entered a new phase of development of nuclear energy. The United States had recently put into operation two new reactor units, while others had come into service in Bulgaria, China, France and Japan. Nonnuclear Belarus, however, could at present provide only about 70 percent of its electricity needs.[28]

Mikhalevich cited the example of neighboring Lithuania, in which the Ignalina nuclear power plant (an RBMK-1500) could provide all that country's basic electricity needs and even export electricity outside its borders, including to Belarus. Ignalina was in fact the basic source of export income of Lithuania. For Belarus, the economic needs were even more acute. Experts from the World Energy Council had published a prognosis that the countries of Eastern Europe would raise their consumption of energy by 1.6–1.7 times by the year 2010. The republican national energy program had anticipated a rise of 1.3, including 1.35 for electricity and 1.45 for heat. It was also hoped to lower the energy cost of the Gross Domestic Product (GDP) by 1.5 to 1.7 times. However, the author observed, this would demand a cardinal reconstruction of energy-consuming branches of the industrial complex. Belarusian stations at present operated on *mazut* (black oil) and natural gas. Neither oil nor coal could be relied on as viable resources for the long term. By the year 2005 a nuclear power station of 1000 megawatts would be required. Given that a nuclear plant takes over four years to build in European countries and over six to seven years in Russia, and that some two to three years of preparatory work would be required in the republic, it was necessary, in Mikhalevich's view, to make a decision on the acceptance of nuclear power at once.[29]

Other scientists have mixed feelings about the development of nuclear power. Academician E.F. Kanaplya, for example, noted that no matter what decision is taken within the republic itself, little can be done to change the fact that Belarus is not merely surrounded by nuclear power stations, but moreover has on three of its borders reactors that are considered unsafe by the majority of international experts.[30] Thus to the northwest is the Ignalina station in Lithuania which, according to one source, is already responsible for the release of low-level radioactivity in this region of the country.[31] On the southern border is Chernobyl itself, which still had two reactors in operation throughout most of 1994, and which the Ukrainian authorities sought to keep in service until the year 2000.[32] To the east was the Smolensk RBMK-1000, the reactors of which constituted the third generation of this ill-fated design. Some 50 kilometers south of Brest Oblast, the Rivne nuclear power plant had three reactors generating electricity. This station was among the most accident-prone in Ukraine. Thus, a cynical observer might maintain that from the safety perspective, whether or not one advocated the construction of a domestic nuclear reactor was somewhat immaterial. Belarus was 'surrounded' by dangerous nuclear power stations beyond its control.

Despite such seemingly rational arguments for the future of nuclear power in the republic, the proponents of this form of energy can only justify their statements if they omit the damage caused to the republic by Chernobyl. If Chernobyl is equated with the new development plans, then the program is no longer financially viable. Two critics, for example, noted that at the International Conference on Ecology in June 1992 in Rio de Janeiro, Shushkevich had declared that the total cost of the damage caused by Chernobyl to the republic was 206 billion rubles, which amounted to 16 annual budgets. Such costs did not include factors such as social-moral costs of the disaster, and the expenses required to procure a calm, stable and healthy way of life for the Belarusian population. How, under such circumstances, the authors inquired, could Shushkevich make the 'puzzling' statement that nevertheless, the republic was to construct its own nuclear power stations in order to alleviate future electricity needs?[33]

In April 1994, the Belarusian Charitable Fund 'For the Children of Chernobyl' under the leadership of Gennadiy Grushevoy, sponsored its second international Congress, 'The World After Chernobyl' in Minsk. Such was the concern over the proposed nuclear power development program in Belarus that a subtitle was appended to the Congress title, translated somewhat ungrammatically into English as 'A World Without Nuclear Menace to Mankind.' Grushevoy pointed out that one of the goals of the Congress was to provide alternatives to nuclear energy development and in his view it succeeded in this aim.[34] The chief spokesperson on this subject was Boris Savitskiy, a parliamentary deputy, and the Chairman of

Belarus

the Permanent Committee on Questions of Ecology and the Rational Use of Natural Resources.

Savitsky acknowledged the energy crisis in Belarus. In 1993, he noted, Belarusian power stations based on gas and oil produced some 33.4 billion kilowatt hours (bkWh) of electricity. The country imported 7.4 bkWh, but exported a further 1.3 bkWh to Pskov and Bryansk Oblasts of Russia, and to Poland. In 1994, it was estimated, Belarus would require 38.5 bkWh of electricity, principally for industries, utilities, transport and street lighting. Belarusian industries were very wasteful with electricity usage, generally using some two to three times that of developed countries on a per capita basis. In general, the country's use of electric power could be termed excessive. However, the authorities had not applied themselves to this issue; no programs were in place to economize on electricity usage as in Germany or the United States. Not only did Belarus need to invest in large-scale electric power production, it also needed energy conservation programs. In Savitsky's view, to embark on nuclear power engineering would only provide corroboration for the myth of cheap energy and would not allow for the modernization of technology in the economy as a whole.[35]

Nuclear power was not a good energy option for Belarus, Savitsky argued. The country had no fuel or waste storage and processing facilities, and even if power stations were constructed, the industry would be dependent on other countries not only for waste disposal, but also for the components and spare parts of the reactor unit itself, since these were not manufactured in Belarus. Specialists would have to be trained outside the republic. Further, a station initiated today (1994) would not be completed until the period 2005–10 and could provide at best for 7 percent of energy needs. In Savitsky's view, it is far more rational to utilize both current and alternate energy sources, and change the whole structure of the power engineering industry. In contrast to the arguments cited above, Savitsky maintained that wind-generated electricity has strong possibilities in Belarus.[36]

Savitsky received support from V.I. Rusan of the Belarusian Research Institute of Energy and Agriculture who enunciated a powerful appeal for the republic to develop an energy program that was not based on nuclear power. In his view, the agro-industrial complex of the nation could take significant steps in energy conservation. First, the existing design of many productive enterprises could be modernized. Second, it would be possible to make use of secondary fuel and power sources either directly or indirectly. Enterprises should begin to ration the consumption of power in addition to using secondary sources and recycling of energy. Like

Savitsky, Rusan believed that the wind resources of the republic are significant and the use of wind turbine electrogenerators for 3000 hours would result in annual energy production of some 20 million kilowatt hours, or a saving of 11–12 million tons of organic fuel. Even solar energy could be used for heating, hot water, drying of agricultural products and the like. Rusan also outlined plans for the use of household garbage to produce electric and thermal power, and the use of organic wastes (such as manure and sewage) to produce gas. Finally, as the republic has long been known as a respository of peat and wood, these resources also might be utilized, in addition to coal, bituminous shale, and oil.[37]

On the question of nuclear power therefore, there are widely opposing views, and some strengths and weaknesses on both sides. Nuclear energy is not a short-term solution to the energy problems of Belarus, though it could provide a more secure guarantee of energy output in the future than some of the ideas postulated by Rusan and Savitsky. Nonetheless, the situation in Belarus seems plain. The population has been traumatized by Chernobyl. The development of nuclear power in such circumstances is something of a non-issue if the government pays heed to the wishes of the public. The debate on nuclear power reached a culmination point in 1993–4, a time when both Russia and Ukraine were anxious to continue nuclear energy development. Ukraine's lifting of its moratorium on the commissioning of new plants in October 1993 and the continuing operation of the Chernobyl station – despite a statement declaring it unsafe from the International Atomic Energy Agency in the spring of 1994 – being cases in point. That such a debate could take place in Belarus is a further sign of retarded thinking at the highest levels of society. In reality, Belarus has neither the technicians nor the funds to embark on a nuclear power program. But the debate is very reminiscent of the 1970s in the Soviet Union, when nuclear power was often seen as the panacea for all energy dilemmas.

THE VIEW FROM 1995

The early months of 1995 brought the republic into the deepest political crisis to date. For the first time in the post-Soviet period, there appeared to be a realistic chance that Belarus might degenerate into a civil war situation. The impetus for the conflict came from the office of the president. In April 1995 during the 12th session of the Belarusian Supreme Soviet, following the signing in Minsk of a Treaty of Friendship and Cooperation with Russia, Lukashenka indicated to the parliament that the May 1995 parliamentary elections should be accompanied by a referendum on four issues: should the Russian language receive the status of a state language equally with Belarusian? should Belarus have a new state flag and national symbols? should Belarus be integrated economically with Russia? and should the president be permitted to amend the Consitution and acquire the authority to dissolve the parliament prior to the end of its term if the latter were shown to be in violation of the Constitution? Aside from the fourth question, which is representative of a direct power struggle between the organs of authority, i.e. the president and the parliament, the referendum questions pose a direct threat to the future of Belarus as an independent state. Indeed some form of national existence without a language, state symbols, a flag, and with Russian authorities policing Belarus's borders (customs integration began on 6 May) and dictating foreign policy appears inconceivable.

Fourteen opposition deputies (including the BPF leader, Zyanon Paz'nyak) responded by organizing a hunger strike around the main podium in parliament. Subsequently the house rejected three out of the four referendum questions (on the Russian language, the state flag and symbols, and the right of the president to dissolve parliament). The strike continued through the night of 11 April before the protesters were forcibly evicted by the militia and presidential guard on the pretext that a bomb threat had been received by telephone. When the parliamentary session resumed the following day, the protesters were not permitted to return, and a cowed parliament accepted all three questions as valid for the 14 May referendum.[38] Apart from the question of state symbols – an issue on which the population is deeply divided – there appeared to be little doubt that the referendum questions would receive positive support from a majority of the population. There was also little indication in Minsk and other cities in April 1995 of any significant public disapproval of the president's dictatorial policies.

What, then, does the future hold for Belarus and its population, including the victims of the Chernobyl disaster? We have argued that the present century, which has brought independence to numerous states as a result of the dissolution of great empires (the British, French, German, Austrian, Russian and Soviet, for example), has been less kind to Belarus. That country is unique in several respects. It is a republic with a majority of citizens who appear to believe that life under the Soviet Union was preferable to present-day existence and who maintain a deep distrust for democratic principles and a market economy. Its president is a provincial politician whose outlook reflects closely that of his compatriots, with the exception of the nationally conscious elements.

Moreover, Lukashenka's intention appears to be to eliminate the Belarusian Popular Front as a political force. The BPF, conversely, might be perceived as the conscience of Belarusians as a nation. By 1995, the official proclamations and public statements of its leaders manifested a deeply-held Russophobia. Its newspaper *Svaboda*, which was being circulated semi-clandestinely in the spring of 1995 following the removal of its editor and a pending court case for libel, has ridiculed the president and has depicted him in cartoons and poems as a modern-day Stalin.[39] The BPF has also identified itself more closely than ever with its leader, Paz'nyak, who has been able to command a solid band of supporters, representing over 10 percent of the voting electorate, but who is also the most intensely disliked of any individual politician in Belarus.[40]

These political events are very pertinent to the central themes of this monograph. The BPF owed its origins to three central issues: the question of Belarusian language and cultural development; the uncovering of Stalinist atrocities in Belarus; and the disaster at Chernobyl. All three issues have been addressed by the present leadership of the country. Lukashenka is attempting to attain personal domination and state unity by putting a halt to the growth and dissemination of Belarusian culture and any form of national state development. There can be little doubt that the elevation of Russian to a state language will reduce dramatically the present emphasis on the use of the native language, whether in schools, higher educational institutions, publishing or mere street signs. Lukashenka is aided in this process by the weak geographical and political basis of the Belarusian culture, by the general lack of awareness of historical background,[41] and by the relatively low level of national consciousness in the republic.

Second, the authorities in general have effectively silenced the process of exposing crimes conducted in the Stalinist era. Given the nature of the exposures and the revelations in other former Soviet republics, this is a considerable achievement. One must take into account, however, the very high degree of public apathy in the republic. The attempts by the BPF and by individual historians to uncover some of the worst excesses of the Stalin period in Belarus have been greeted with a deafening silence, or by covert hostility on the part of the general public. The suppression of public information has been enhanced by the continued propagation of the achievements of the Soviet people during the war, and the Belarusian-based partisans in particular. In short, those in authority who have links with the past Stalinist era are anxious to prevent further details from coming to light.

The third key issue, that of Chernobyl, remains unresolved. Lukashenka has not ignored this question. He has given public voice to his concern for child victims of Chernobyl, and in March 1995 he visited the contaminated regions amid considerable publicity.[42] It is difficult to ascertain, however, whether his statements and peregrinations constitute more than ritual. For the victims of Chernobyl – evacuees, liquidators, families who have been resettled or remain in contaminated regions – the future is rendered more uncertain by the virtual eclipse of the parliamentary opposition who, in the past, have been the most effective advocates of their cause. Further, the strongest charitable association, the Belarusian Charitable Fund 'For the Children of Chernobyl,' had been obliged to reregister itself with the authorities and compelled to divide itself into geographical sections early in 1995. At that point, it was too early to tell whether the Fund had been weakened significantly by such decentralization.

Early in 1995, two surveys were conducted that made reference to Chernobyl and allow one to deduce its relative significance as a key issue for families in the republic. The first survey was conducted among 1018 respondents from diverse regions of the republic, and asked the public to cite problems causing the most anxiety. In order, the list was as follows: rising prices; the drop in the living standards of the population; the dissolution of the Soviet Union; the increase in crime; the consequences of Chernobyl; and unemployment.[43] In this list therefore, nine years after the accident, Chernobyl ranked fifth as an issue of concern. Significantly, national existence and the question of language did not feature on the list. The second survey was exclusively about Chernobyl, and divided respondents into three groups: a control group based in Vitsebsk region; those living in the contaminated zone; and those resettled from the zone affected by the accident. The results demonstrated that even for those living the furthest distance from the affected regions, Chernobyl remains an issue of the highest significance. Hence of the Vitsebsk group, 10 percent personally considered Chernobyl the chief issue of the day, while 35.8 percent considered it an issue alongside others. Figures for the resettlers were 17.3 percent and 48.7 percent respectively; and for those in the zone, 20.1 percent and 41.8 percent. Conversely, in the control group, only 6.6 percent considered that Chernobyl was no longer a matter of importance, 1.8 percent of those in the zone and none among the evacuees.[44] If not the principal problem for Belarusians, Chernobyl remains a cause for great concern.

In this volume we have used the word 'catastrophe,' while emphasizing that one must take into account psychological aspects of the medical consequences alongside authentic radiation-induced illnesses. Is this word necessary? One can only answer in the affirmative. Even had there been no medical repercussions from this event and no casualties, it would remain a disaster on a world scale. In Belarus, the problems have been augmented by what we have described as a flawed and destructive national state development in the Soviet era and beyond. Chernobyl has affected the lives of hundreds of thousands of Belarusians. It has entailed a fundamental demographic change: the depopulation of an entire region of a republic, in addition to the damage to agricultural land and forest caused by long-living radioisotopes. Chernobyl did not bring about the collapse of the Soviet regime, nor did it alter fundamentally the political spectrum in the country, but the ramifications of the disaster were nonetheless drastically affected by the political changes and the concomitant economic crisis that has beset Belarus, like other countries. That crisis is likely to continue and to grow. In the spring of 1995, there was little evidence of either economic or political reform in the country. The currency remained relatively stable against the dollar only as a result of the employment of state reserves to support the Belarusian ruble. There was every indication of an imminent and devastating economic collapse within a 6–12 month period. In the interim, an unpredictable and ruthless president was consolidating his power. Taken together, the consequences of Chernobyl – medical and social – and the political events constitute a catastrophe for the victims of Chernobyl. There will not be a casualty list in the future to match that at Bhopal or the 1988 Armenian earthquake, yet the nuclear disaster continues to have an impact on the lives of families in at least one-third of this nuclear-power-free republic, to say nothing of future generations.

EPILOGUE

On 9 May 1995, the fiftieth anniversary of the Soviet victory over Nazi Germany was celebrated in Minsk, as elsewhere in the former Soviet Union. In the Belarusian capital, however, the occasion was more than ceremonial; it signified a possible path for the future. At the demonstration in Independence Square, those carrying old Soviet flags greatly outnumbered those bearing the national flag of Belarus. President Lukashenka's speech recalled the 'glorious days' of the Soviet past, when the borders of the country stretched to the Pacific. The underlying message of these words could hardly be mistaken. The leader of Belarus was expressing his desire, and ostensibly that of his people, to 'return' to the Russian fold. Ironically the disastrous events of Chernobyl, which proved to be significant in the dissolution of the Soviet Union (and thereby the bond of Belarus with Russia), were also likely to be a factor in the collapse of Belarus as an independent state. The fifty-year anniversary reminded the demonstrators of the time when Belarus had been rescued from the east at a critical juncture in its history. Was it about to happen again? And what of the Russians? Would they be prepared to take on the additional burden of the struggling republic that had failed to embark on significant economic reforms?

It remains to be said that there is an alternative route for Belarus, namely the long process of democratic development that must be accompanied by increased self-awareness of the events of the recent past. If Belarus enjoyed a golden age of culture in the 1920s, and if today a majority of children are studying the Belarusian language in school, then such progress has to be attributed to the efforts of the nationally conscious democrats, many of whom are embraced by the Belarusian Popular Front. Belarusians are in general highly educated and adaptable to difficult conditions. The impact of the Soviet regime on this republic has yet to be revealed fully to the population, however, and the belief that a return to Russia would bring about economic prosperity and progress is naive in the extreme. One only has to look at the decline and degradation of the Russian Far East to perceive how Moscow has neglected areas outside the immediate purview of its political leaders.

On the other hand, the BPF must, in turn, recognize that 'instant' change is unattainable; that the population has not yet emerged from the Soviet chrysalis; and that any attempt at forced 'Belarusification' would only alienate further a majority of the population. If national nihilism is to be overcome; if Belarus as an independent state is to have a future; and if a new generation is to emerge from the disaster caused by Chernobyl with renewed hope; then the attitude of the national intelligentsia must change also. Russian-speaking Belarusians in Vitsebsk may as yet have little in common with Belarusian speakers in Brest region. They cannot, however, be belittled or admonished for either their lack of national consciousness or their failure to adopt their native language after 70 years of non-usage. The same applies to ethnic Russians within Belarus who have to be treated as equal citizens in the new republic. With tolerance and patience, then, Belarusians may learn to recognize the salient fact: that from the East Slavic nations of the past has emerged not one, but three states, two of which are Russia and Ukraine. The third, Belarus, at present lacks the population, international prestige and political security of the first two. Yet it is by no means an anomaly created by the collapse of an empire. Historically and morally, it has an equal right to exist. What is lacking as the twentieth century draws to a close is the will to survive alone.

Notes

1 SOVIET RULE: REPRESSION AND URBANIZATION

1. The term Belarus is used throughout to denote the Belarusian provinces of the Russian Empire, the regions of Belarus incorporated into Poland after World War I, the Belorussian (Byelorussian) Soviet Socialist Republic and its successor, the independent Republic of Belarus.
2. These and other aspects of the post-Soviet era are discussed in detail in Chapter 5.
3. *Britannica Book of the Year 1994* (Chicago, 1994), p. 562.
4. Borodina et al., 1972, pp. 35–6.
5. See, for example, Smolich, 1993, pp. 2–3.
6. Borodina et al., 1972, p. 95; Pokshishevsky, 1974, p. 159.
7. Reserves of peat are reportedly about 5 billion tons in terms of air-dry mass. Pokshishevsky, 1974, p. 157.
8. Guthier, 1977a, 1977b; Vakar, 1956.
9. Zaprudnik, 1993. The previous monograph was Lubachko, 1972.
10. Institute of Art, Ethnography and Folklore, Academy of Sciences of the BSSR, 1979, pp. 13–14, 18.
11. Ibid., pp. 15, 18.
12. Oddly, however, Minsk was among the last of the 'hero cities' to be created, in 1974.
13. In the historical archives in Minsk, an entire section is comprised of partisan records, which are preserved in minutest detail and constitute a valuable wartime record that has yet to be fully exploited by historians. Access is not restricted, though at the present time, foreigners are expected to negotiate a hard currency fee for their perusal.
14. For two recent biographies of Masherau, see Yakutov, 1992; and Antonovych, 1993.
15. Information of Professor Dmitriy D. Kozikis of the Minsk State Linguistic University who was present in Minsk at that time.
16. Cited in Solchanyk, 8 October 1978.
17. Ignatouski, 1992.
18. Kipel and Kipel, 1988, pp. 323–4.
19. Information provided in Malashko, 1988, p. 68. Neither of these works was available in Minsk bookstores in 1992–4. The only available serious historical monograph on Belarus was that of Ignatouski.
20. The historians included the following (the names are rendered in their Russian form as in the text): Vladimir Orlov, Vol'ga Bobrovka, Mykhas' Bich, Yanka Voynich, Vintsuk Vechorka, Mikola Ermolovich, Ales' Zhlutka, Edvard Zaykovskiy, Emmanuil Ioffe, Vladimir Kazberuk, Yuras' Kalachinskiy, Ibragim Kanapatskiy, Ales' Kratsevich, Leonid Lych and others, and was prepared for publication by Ernest Yalugin and Yazep Yanushkevich.
21. *Narodnaya hazeta*, 27 May 1993, p. 6.

22. Golenchenko and Osmolovskiy, 1993, p. 176.
23. The preliminary program for the Fifth World Congress of Slavic and East European Studies, to be held in Warsaw in August 1995, contained only two panels relating specifically to Belarus, and only one Belarusian scholar was listed on the program, despite the proximity of the location to the republic. There are nonetheless some noted specialists on the Belarusian language, such as Peter Mayo of the University of Sheffield and James Dingley of the School of Slavonic and East European Studies, University of London.
24. Guthier, 1977a, p. 45.
25. Borodina et al., 1972, p. 62.
26. Zaprudnik, 1993, p. 61.
27. Kasperovich, 1985, p. 30.
28. Academy of Sciences of the Belorussian SSR. Institute of History, 1977, pp. 165–6.
29. Guthier, 1977a, p. 43. Another source provides a figure of 17.7 percent Belarusians among the 'large-scale trading–industrial bourgeoisie.' See I.N. Braim, 'Sotsial'noe razvitie gorodov,' in Bandarchik, 1980, p. 31.
30. Kasperovich, 1985, p. 27.
31. I.N. Braim, 'Urbanizatsiya i nekotorye izmeneniya v demograficheskoy strukture,' in Bandarchik, 1980, p. 48.
32. Kasperovich, 1985, p. 27.
33. See, for example, ibid., p. 32.
34. Notably, however, many of the symbols of the present independent state originated at this time, including the seal – a knight on horseback – formerly used by the Grand Duchy of Lithuania – and the white-red-white horizontally striped flag. See, for example, Kipel and Kipel, 1988, p. 390.
35. Zaprudnik, 1993, p. 70. It consisted at this time only of the districts of Minsk, Hrodna, Mahileu and Smolensk.
36. Mienski, 1988, p. 165.
37. See, for example, ibid., pp. 167–70. Urban, 1988, p. 192, notes that this new BSSR had an area of 52 316 square kilometers and a population of 1.5 million, whereas the ethnic Belarusian territories consisted of more than 317 000 square kilometers and about 12 million people. the Treaty of Riga, he notes, had ceded some 108 000 square kilometers of Belarusian territory and four million people to Poland, while a smaller area with 300 000 population was handed over to Latvia. Therefore the Russians themselves expropriated some 150 000 square kilometers of land, with an estimated population of six million.
38. Urban, 1988, pp. 193–4, states that the agreement of early 1924 between the Russian Federation and the BSSR saw the transfer to the latter of a substantial territory in the Vitsebsk, Mahileu and Homel' regions, in addition to three small rural regions of the Mstislavl part of Smolensk Oblast. As a result, the BSSR increased in size from 52 316 to 109 764 square kilometers and the population from 1.5 to 4.2 million. In 1926 after a so-called popular plebiscite, the remainder of Homel' Oblast was annexed to the BSSR by a decree of 6 December. The territory then comprised 125 522 square kilometers and the population was increased to 5 million people. Urban attributes this concession of territory in part to the agitation of the Belarusian movement in Poland. There was a further slight addition of territory from Smolensk Oblast in 1961

and 1964. Paz'nyak attributes the transfer of these lands to the demands of the local Smolensk Belarusian population. Paz'nyak, 1992, p. 191.

39. Zaprudnik, 1993, p. 78.
40. Lubachko, 1972, p. 85.
41. Borodina et al., 1972, p. 65.
42. Guthier, 1977a, p. 60.
43. Symon Kabysh, 'Genocide of the Belorussians,' in Kipel and Kipel, 1988, pp. 229–41.
44. The author has recently examined the issues of Kuropaty and its various interpretations. See Marples, 1994b, pp. 513–23.
45. Paz'nyak and Shmyhaleu, 1988; see also Poznyak (Paz'nyak), 1992, p. 16.
46. Marples, 1994b.
47. Kuznetsov, 1995, p. 2.
48. Zaprudnik, 1993, p. 85.
49. Dementeya, 1989, p. 2.
50. Ibid.
51. Symon Kabysh, in Kipel and Kipel, 1988, p. 237. Zaprudnik provides a figure of 300 000 total deportations from western Belarus prior to the German invasion. Zaprudnik, 1993, p. 91.
52. *Istoriya Velikoy Otechestvennoy voiny*, Vol. 1, p. 249.
53. It is instructive to compare Iwanow's comments with those of a Soviet version of the annexation of Western Belarus:

> An historic event took place in the life of the Byelorussian people in September 1939, when their dream of reunification for all the Byelorussian lands into a Byelorussian Soviet state came true .... On November 2, the Supreme Soviet of the USSR passed a law incorporating Western Byelorussia into the USSR and reuniting it with the rest of Byelorussia into a single Byelorussian Soviet Socialist state.

Borodina, et al., 1972, p. 66. The authors were evidently far from nonplussed by the noncorrelation of historic dream and Soviet state. Moreover, it was important at all times from the Soviet perspective to emphasize the role of the Moscow authorities in realizing alleged ambitions of a subject people.

54. Iwanow, 1991, pp. 255–6.
55. Dementeya, 1989, p. 2.
56. Why was Vilna transferred to Lithuania? According to one historian, Stalin's intention was to compensate Lithuanians for being forced to accept Soviet military garrisons on their territory. Vardys, 1991, p. 269.
57. 'Voprosov i otvetov iz istorii Belarusi,' *Narodnaya hazeta*, 1 June 1993, p. 3.
58. Kravchenko and Marchenko, 1965, p. 49.
59. Lubachko, 1972, p. 162. There were various other German concessions to nationalist Belarusians at the latter end of the war, including the formation of a Belarusian Defense Committee and the convocation of the Second All-Belarusian Congress on 27 June 1944, attended by over 1000 delegates. The latter were forced to flee with the retreating Germans within a few days. See Lubachko, 1972, p. 163.

60. *Istoriya Velikoy Otechestvennoy voiny*, Vol. 3, p. 439.
61. According to one Soviet source, by the end of July 1941, just one month after the German invasion of Soviet territory, 231 partisan units and groups operated on Belarusian territory, with 12 000 participants. Academy of Sciences of the Belorussian SSR, Institute of History, Vol. 2, 1961, pp. 434–5. Such figures are clearly exaggerated.
62. Ibid., p. 460. Another source provides a total of 440 000 partisans, compared to one million Belarusians who were fighting in the Red Army. Borodina et al., 1972, p. 66. Again the figure appears inflated.
63. Lubachko, 1972, p. 150.
64. According to information provided in the Museum of the Great Patriotic War in Minsk.
65. Kravchenko and Marchenko, 1965, pp. 60–1.
66. *Istoriya Velikoy Otechestvennoy voiny*, Vol. 4, pp. 152–3.
67. Academy of Sciences of the Belorussian SSR. Institute of History, 1977, p. 422.
68. Ibid., Vol. 5, p. 431–2.
69. Over 300 000 Red Army soldiers from Belarus and over 100 000 Belarusian partisans were decorated for bravery during the Great Patriotic War; 295 soldiers and 74 partisans received the award of Hero of the Soviet Union. Academy of Sciences of the Belorussian SSR. Institute of History, 1961, p. 494.
70. Academy of Sciences of the Belorussian SSR. Institute of History, 1961, p. 510.
71. Concerning the immediate postwar years, Martin McCauley notes that:

    Down on the farm it was quite a different tale. It was the return of the bad old days of the 1930s: no incentives, the centralisation of every decision which could be centralised, a harsh paternalistic attitude towards the rural sector, with farms regarded as the milch cows of the cities and industry. McCauley, 1993, p. 187.

72. Academy of Sciences of the Belorussian SSR. Institute of History, 1961, p. 536.
73. Even the first postwar Five-Year Plan (1946–50) began heavy concentration upon industrial development and virtually ignored agriculture. At that time was initiated the manufacture of automobiles, tractors, locomotives, televisions, watches, trucks and motorcycles. Kiselev, 1964, p. 21.
74. Kharevskiy, 1990, pp. 42–3.
75. State Committee of the Belorussian SSR for Statistics, 1990, p. 14.
76. Gol'bin, 1988, p. 97.
77. Central Statistical Administration with the BSSR Council of Ministers, 1970, p. 9.

2 LANGUAGE AND CULTURE: NATIONAL NIHILISM?

1. Kasperovich, 1985, pp. 21–4.
2. *Kommunist Belorussii*, No. 8 (1991): 34.

3. Ibid., p. 30.
4. Danilov, 1993, p. 60.
5. Shakhot'ko, 1975, pp. 139, 147–8.
6. State Committee of the Belorussian SSR for Statistics, 1990, p. 18.
7. Figures provided for the author by Sergey Laptev, a member of the Minsk City Council.
8. Zaprudnik, 1993, p. 3.
9. Guthier, 1977a, p. 46.
10. Shabailov, 1989, pp. 271–2.
11. *Belarus News*, No. 6, April 1995, p. 8.
12. Eshin, 1970, p. 30.
13. Mikhnevich, 1985, p. 19.
14. Lubachko, 1972, pp. 80–1.
15. Academy of Sciences of the Belorussian SSR, Institute of History, 1977, p. 288.
16. *Belarus News*, No. 6, April 1995, p. 8.
17. Academy of Sciences of the Belorussian SSR. Institute of History, 1979.
18. Lubachko, 1972, pp. 170–2.
19. Thus prior to the October Revolution in the territories of Belarus, less than 1500 people attended higher educational institutions. In the 1966–7 educational year the figure was almost 250 000. Eshin, 1970, p. 245.
20. Kravchenko and Marchenko, 1965, pp. 66–7.
21. Sobolenko in Bandarchik, 1980, p. 196.
22. Ibid., p. 203.
23. A complete population breakdown by ethnic origin has been published for the subsequent 1979 census for the city of Minsk. It indicates that Belarusians made up 68.7 percent of the total population (864 510) people; Russians 22 percent (277 166), followed by Jews at 3.7 percent, Ukrainians at 3.6 percent and Poles at 1.1 percent. *Kommunist Belorussii*, No. 9, 1988, p. 44. One can see therefore that most ethnic Belarusians in the capital city cited Russian as their mother tongue.
24. Guthier, 1977b, p. 275.
25. Solchanyk, 1979, p. 5.
26. Mikhnevich, 1985, p. 18.
27. Volodin, 1979, pp. 39, 41.
28. Paz'nyak, 1988a.
29. Guthier, 1977b, p. 275.
30. Central Statistical Administration of the Belorussian SSR, 1970, p. 309.
31. Paz'nyak, 1988a.
32. *Grodnenskaya pravda*, 30 October 1990.
33. Mal'dis, 1989, p. 73.
34. Ibid., p. 74.
35. That is, the Brezhnev period. The so-called period of stagnation is often used to denote the entire Brezhnev period, 1964–82. More accurately, it applies to the period 1974–82, when the worst abuses of the Brezhnev era occurred.
36. Mal'dis, 1989, pp. 74–6.
37. Malashko, 1988, p. 67.
38. Ibid., p. 68.

39. Rutskaya, 1989, p. 77.
40. Trusov, 1989, p. 15.
41. See, for example, *Sovetskaya Belorussiya*, 25 September 1989, p. 1; *Respublika*, 12 April 1995, p. 2; and Padluzhny, 1993, p. 77.
42. *Sovetskaya Belorussiya*, 25 September 1990.
43. Belarusian Popular Front '*Adradzhennie*', 30 May 1993.
44. This may be presumptuous. In the past, Communists in Belarus have on occasion embraced the language issue. By and large, however, it has been the BPF that has pursued the issue as an essential ingredient of its policies. In the first six months of his period of office as president, Lukashenka delivered all his speeches in Russian. As noted in Chapter 5, by the spring of 1995 he was seeking to make Russian a state language in Belarus.
45. Smolenskiy, 1989, p. 16.
46. Belarusian Sociological Service 'Public Opinion', 1992, pp. 54, 56.
47. Ibid., pp. 60–1.
48. See, for example, *Svaboda*, 7 April 1995, p. 1; Yanovich, 1995, p. 2; and *Svabodnyya prafsayuzy*, 12 April 1995, p. 1.
49. See, for example, *Respublika*, 12 April 1995, p. 4.
50. *Minsk Economic News*, No. 6, March 1995, p. 4.
51. Kozikis, 1995.
52. The settlement of Vetka, in Homel' Oblast, was also at one time a center of Old Believers in the Russian Empire. The museum there today contains artworks, icons and other materials from that period. One could assert therefore that Chernobyl not only affected the repositories of specifically Belarusian culture, but the history of religious belief in the Russian Empire as a whole.
53. The noted poet Mikola Myalitski established a foundation called 'Paleski smutak' in order to provide spiritual support to thousands of Chernobyl victims and to preserve the historical memory of the abandoned villages, 'to immortalize them in literature, art, and cinematic works.' Myalitski's fear is that the nuclear zone will become a wasteland, and that the cultural background of this area of the republic will be forgotten. Myalitski, 8 October 1993.

3   THE CONTAMINATED ZONE

1. *Chernobyl'skaya katastrofa*, 1992, p. 158.
2. *Sovetskaya Belorussiya*, 28 March 1995, p. 1.
3. Kartel, 1991, p. 12.
4. Udovenko, 1991, p. 20. Some sources give a larger area of fallout. For example, one source declares that 28 raions of Belarus were contaminated by Chernobyl fallout, totalling 46 500 square kilometers of territory. *Dobryy vecher*, 24 February 1993.
5. Kazakov, et al., 1991, p. 4.
6. Gurtovenko, 1995, p. 1.
7. This area was evidently discovered in the period January 1988 to mid-1989, and also embraced the western oblast of Hrodna. Demichev, p. 28.

8. As shown on the detailed map: Committee of Land Measurement, 1992.
9. Kryzhanovskiy, 1990, p. 50. See also *Chernobyl'skaya katastrofa*, 1992, pp. 19–20.
10. Kadatskiy et al., 1991, p. 27.
11. Kozlovskaya, 1991, p. 28.
12. Kartel, 1991, p. 12.
13. I have dealt extensively with the fallout in Ukraine in Marples, 1991.
14. The extent of the fallout in Russia is often underestimated, mainly because of the size of that country. Nonetheless, the fallout was substantial, affecting 14 oblasts with a total population of 2.34 million people and 50,000 square kilometers of territory. The oblasts include, in addition to Bryansk, St. Petersburg, Tambov and Smolensk, in addition to the Mordovian Autonomous Republic. Bryansk Oblast is the most severely contaminated. It possesses an area of 2130 square kilometers with over 5 curies/km^2 of cesium, and a small region of 240 kilometers with over 40 curies. In Bryansk also there are appreciable amounts of strontium-90 in the soil. *Meditsinskaya gazeta*, No. 98, December 11, 1992, pp. 1–6; and *Gosudarstvenniy doklad*, 1992, p. 32.
15. Kanaplya, 1992.
16. This figure clearly includes those territories in the initial zone of evacuation around the reactor; otherwise it contradicts the information provided above that Ukraine does not possess settlements with such high levels of radiation in the soil.
17. *Holas Vetkaushchyny*, 26 October 1993.
18. Kartel, 1991, pp. 14–15; Petrayev and Leynora, 1991, p. 21; Marples, 1992, pp. 421, 423.
19. Demidchik, 1993a.
20. *Chernobyl'skaya katastrofa*, 1992, p. 15.
21. Ibid., p. 16; and Marples, 1994a, p. 103. Evidently this law was modeled on the one previously adopted by the Supreme Soviet of Ukraine.
22. Demichev, 1993, p. 29.
23. *Gosudarstvennaya programma*, 1989.
24. 'Zakon "O sotsial'noy zashchite grazhdan, postradavshikh ot katastrofy na Chernobyl'skoy AES" i khod ego realizatsii', in *Chernobyl'skaya katastrofa*, 1992, pp. 192–4.
25. 'Zakon "O pravovom rezhime territoriy, podvergshikhsya radioaktivnomu zagryazneniyu v rezul'tate katastrofy na Chernobyl'skoy AES," cited in *Chernobyl'skaya katastrofa*, 1992, pp. 195–6.
26. The full title is cumbersome: the State Committee of the Republic of Belarus on the Problems Arising from the Consequences of the Catastrophe at the Chernobyl Nuclear Power Plant. It has recently been transformed into a ministry.
27. *Chernobyl'skaya katastrofa*, 1992, p. 197.
28. Ibid., pp. 147–8.
29. IAC, 1991, p. 39.
30. *Zvyazda*, 13 August 1993.
31. See, for example, Zverev and Grushevoy, 1992, which is discussed in more detail below.
32. *Narodnaya hazeta*, October 2, 1993.

33. Thus Pyatraeu himself declared that there was no need for interference in any area in which the accumulated individual exposure dose was less than a millisievert, since such a criterion would fully conform to international standards. He also acknowledged, however, that part of the Belarusian population was living in contaminated zones in which the annual equivalent dose was less than 1mSv. Petrayev, 1993. Would not this exclude from concern also, however, areas that may be outside the official contamination zone in which the exposure dose level exceeds 1mSv? And how would it take into account the amount of exposure to which the population had already been exposed in the seven (at that time) post-Chernobyl years? Such questions are not elucidated in the revised concept.

34. *Zvyazda*, 13 August 1993.

35. *Dobryy vecher*, 5 September 1993.

36. *Golos Krasnopol'shchyny*, 11 September 1993.

37. *Gomel'skaya pravda*, 3 December 1992 and ff.

38. In the summer of 1992, the Belarusian Supreme Soviet issued a new law, 'Concerning the protection of the surrounding environment,' which highlighted the key predicaments facing the republic in the ecological sphere. See *Ekologiya Minska*, Nos 14–15, August 1992, pp. 3–6.

39. The figures cited in the article are 117 million distributed from an assigned 129 million rubles. By the prices of 1992–3, such totals would have been grossly inadequate. In September 1992, Georgiy Lepin of the Belarusian Union of Chernobyl', affiliated with the Ministry of Internal Affairs, commented on the proposed construction of sanatoria for children suffering from the effects of Chernobyl. One such sanatorium was said to cost 25 million German deutschmarks, and the Chernobyl Union was declared to be reliant on German aid for its construction since that figure was well beyond its means. See *Ekho Chernobylya*, Nos 29–32, September 1992, p. 2.

40. The author has on two occasions asked about the comprehensiveness of such registers. At both the Children's Hospital No. 3 (by the matron) in December 1993 and at the Institute of Radiobiology (by its director, Evgeniy Kanaplya) in April 1993, he was informed that the list is complete and that the scientific and medical authorities have a record of each person in the republic affected by Chernobyl radiation. However, I have never been shown such a list, despite repeated requests.

41. Initially, Grushevoy was a member of the Belarusian Popular Front, but has had some differences with its leader, Zyanon Paz'nyak. In the 1994 elections for the presidency, he appeared to be lending support to his former adversary, Stanislau Shushkevich. As he is an ethnic Russian, his name has been transliterated in the Russian form.

42. Interview with Gennadiy and Irina Grushevoy, Minsk, October 13, 1992. One of the more pertinent points raised during this interview was the fact that those families outside the country that wish to sponsor children from the contaminated regions for periods of recuperation must send individual letters of invitation to the Belarusian Ministry of Foreign Affairs. This stipulation had greatly increased the time needed to finalize accommodation for children with their host families (see below for more information on nongovernmental aid).

43. This is an abbreviated translation of the title of the program, which is rendered fully in English as 'The National Program of Prophylactic Genetic Consequences arising from the accident at the Chernobyl nuclear power station.'

44. *Narodnaya hazeta*, 16 December 1992 and ff.

45. Bebenin, 18 November 1994.

46. IAC, 1991, p. 44.

47. Institute of Radiation Medicine [IRM], 1992, p. 2.

48. *Pravda*, 28 February 1991.

49. Kryzhanovskiy, 1990, p. 50.

50. Shumarova, 1991, p. 64. It is also notable, nonetheless, that while the local population was anxious to be moved, other Belarusian residents were moving into Brahin, attracted by the double wages offered, in addition to what was termed a 'coffin bonus' of 30 rubles per month. Shumarova, 1991, p. 64.

51. Demichev, 1993, p. 29.

52. Kanaplya, 14 April 1993.

53. *Katastrofa*, 24 April 1993. The newspaper is an occasional publication of the influential Minsk daily *Respublika*.

54. Figures supplied by Sergey Laptev, a deputy of the Minsk City Council, in 1994.

55. The plan was to resettle 59 families with 104 people.

56. The plan anticipated the removal of one family with two members.

57. *Katastrofa*, 24 April 1993.

58. *Respublika*, 20 July 1993.

59. Ibid.

60. Ibid.

61. *Mahileuska prauda*, 28 August 1993.

62. Demichev, 1993, p. 28.

63. *Zvyazda*, 4 April 1993.

64. *Katastrofa*, 24 April 1993.

65. *Katastrofa*, 11 August 1993.

66. Semkin, 1994, p. 22.

67. *Vitebskiy rabochiy*, 6 February 1991.

68. Semkin, 1994, p. 22.

69. *Chacherski vesnik*, 24 July 1993.

70. *Golos Krasnopol'shchyny*, 11 September 1993.

71. The figures are derived from two separate articles. The first, which does not concentrate on numbers specifically, mentions that there were 12 000 'Chernobylites' living in Minsk: *Dobryy vechar*, 20 September 1993. The second comments that 25 000 resettlers were living in the city, including 9000 children, and that most of the adults did not have jobs: *Vecherniy Minsk*, 22 June 1994.

72. *Zvyazda*, 4 September 1993.

73. It has been alleged by several sources that increased radiation background as a result of the Chernobyl disaster lowered the resistance of the body generally and rendered the victims more readily susceptible to any diseases.

74. Minsk City Council, Department of Health, 1993.

75. David F. Duke, a Canadian doctoral candidate in history at the University of Alberta, who was teaching at that time at Minsk State Linguistic University,

visited Malinauka-4 in November 1994, and was informed by a local doctor, Tatyana Abramchuk, that the prevailing wind usually blows the pollution over the apartment blocks. A build-up of deposits was already in evidence and was eating away at the concrete.

76. A.N. Stozharov, 14 December 1993.
77. Demichev, 1993, p. 28.
78. Petrayev, 23 November 1993.
79. *Katastrofa*, 11 August 1993.
80. Ibid.
81. *Dobryy vecher*, 4 December 1992.
82. Yakushev, 1992.
83. *Golos Krasnopol'shchyny*, 14 July 1993.
84. *Mayak prydnyapravya*, 21 July 1993.
85. *Katastrofa*, 24 April 1993.
86. Semkin, 1994, pp. 22–3.
87. *Katastrofa*, 24 April 1993.
88. Ibid.
89. *Katastrofa*, 11 August 1993.
90. Ibid.
91. Ibid.
92. Kalynovich, 1995, p. 5.
93. This village is not to be confused with the town of Vileyka, in Minsk Oblast, which was discussed in Chapter 1.
94. Could these comments be applied to Belarusian villages generally, i.e., in areas that were not contaminated by Chernobyl? Certainly the roads generally are poor and living conditions well below those offered in urban centers. However, all the villages I visited in the republic outside the contaminated zone were in much better condition than those in the Chavusy region. One may be on more shaky ground in attempting to analyze the relative consumption of vodka. In the spring of 1995, and indeed over the three years encompassed by this study, such consumption appeared to be universally heavy in all settlements, but particularly in rural areas. This subject is worthy of serious study.
95. In truth, it seems highly unlikely that the thyroid gland problems here were a result of the spread of radiation. Such ailments are generally associated – insofar as they apply to Chernobyl-related illnesses – with the spread of radioactive iodine. Such iodine would no longer have been a factor after the summer of 1986 because of the very short half-life (8 days) of iodine-131.
96. *Katastrofa*, 24 April 1993.
97. For example, a dust storm in the southwestern raions of Homel' Oblast in February 1993 evidently 'violated the ecological balance.' At this time of year, especially when the winter had been snowless, there was often an increased migration of radioactive particles, which could be transported long distances from their original burial site, some 3–5 centimeters deep in the soil. *Sovetskaya Belorussiya*, 23 February 1993.
98. *Inter alia*, the authorities have sometimes held up transportation of medical and other provisions by truck at the western border of the country, sometimes for days at a time. Information of Burkhard Homeyer, Chairman of the Children of Chernobyl organization in Germany, April 1994. See also Homeyer, 1994, p. 99.

99. *Dobryy vecher*, 24 August 1993.
100. *Zvyazda*, 4 September 1993.
101. *Nastaunitskaya hazeta*, 4 August 1993.
102. Cited in *Narodnaya hazeta*, April 16, 1992.
103. Lupach, 1992, p. 112.
104. About 70 percent of the Fund's external funding comes from Germany. Grushevoy himself is a fluent German speaker with excellent contacts in that country. Germany, noted Grushevoy, was indisputably the leader in the realization of the Fund's program. Lupach, 1992, p. 113.
105. This interview took place in Minsk on 20 April 1993.
106. Ibid., and ff.
107. Lupach, 1992, p. 114.
108. Grushevoy, 11 December 1993.
109. Such disputes are not unusual in Minsk and are often short-lived. One can attribute them, at least in part, to the atmosphere of tension that exists both as a result of Chernobyl directly, and as a result of the unfriendly relations between the government and some of the charitable organizations.
110. Duss, 1992.
111. *Vecherniy Minsk*, 22 June 1994.
112. *Sovetskaya Belorussiya*, 1 July 1994; *Meditsinskiy vestnik*, 28 July 1994.
113. *Respublika*, 19 February 1993.
114 *Gomel'skaya pravda*, 7 September 1993.
115. *Sovetskaya Belorussiya*, 2 April 1993 and ff.
116. See, for example, *Respublika*, 22 July 1994.
117. Occasionally foreign participants in these programs have been attacked by government and parliamentary leaders. In August 1993, Professor Edmund Lengfelder of Munich University, the Chairman of the Otto Hugo Radiological Institute, successfully sued Inesa Drabysheuskaya, a parliamentary deputy from Homel' region, and the newspaper *Homel'skye vedamosty*, for 'distorting the truth' and humiliating him in a published article. The newspaper was forced to pay 500 000 rubles in damages and 97 017 rubles in legal expenses, and was obliged to publish an apology. However, 60 parliamentary deputies then published articles in several newspapers attacking the German scientist (whether individually or collectively is not specified). Several medical institutions and charitable organizations rushed to the professor's defense pointing out his extensive aid to sick children, his resolving of visa problems for a group of Belarusian children travelling to Belgium, and other examples. *Zvyazda*, 9 September 1993. The incident is not unique and reflects official resentment at the involvement of foreign agencies in the affairs of Belarus.
118. Grushevoy, 20 April 1993.
119. *Ekho Chernobylya*, No. 29–32, September 1992, p. 2.

4   MEDICAL CONSEQUENCES OF A NUCLEAR DISASTER

1. IRM, 1992, p. 2.
2. In 1992, an Endrocrinological Center of Children and Adolescents was founded in Minsk by various public organizations with the stated goal of examining the influence of large city industrial pollution on the state of

health of children and young people. A seven-year monitoring program was established (1993–2000) under its director, N.A. Gres'. Initially, the Center compared the health of children in different zones of the city, by analyzing the amount of lead, mercury, zinc and copper from samples of urine. Some clear conclusions emerged. First, the most industrialized region (Zavadskiy district) revealed more serious health problems among children that the relatively clean area (Zelyoniy Luh region on the periphery of the city). Second, an examination of 3150 children aged three months to 17 years indicated that over 76 per cent of them had concentrations of nitrates in their urine that considerably exceeded the permissible norms. Minsk City Council, 1993, pp. 18, 62. In 1994 the Center, together with the health department of the Minsk City Council produced a book on ecology and the state of health of Minsk children after the Chernobyl disaster. Minsk City Council, 1994.

3. It is 1440 rubles per month, a sum that by the end of 1994 would have bought in hard currency exchange about US15¢. *Mogilevskie vedomosti*, 5 November 1994.

4. The figure is a very low one. Compare other prognostications of future cancer deaths related to Chernobyl in Marples, 1988, Chapter 1.

5. Cited in *Respublika*, 12 May 1993.

6. Krauchanka, 1990. A draft copy of this speech was made available to me.

7. The uppermost region of the throat located behind the nose. See Anderson, 1994, p. 1047.

8. Krauchanka, 1990, p. 7.

9. Ibid., p. 9.

10. *Vecherniy Minsk*, 31 January 1991.

11. Impairment of muscle tone, and often involving the head, neck, and tongue. Anderson, 1994, p. 517.

12. Anderson, 1994, p. 331, defines chronic as '[a disease or disorder] developing slowly and persisting for a long period of time, often for the remainder of the lifetime of the individual.'

13. Kazakov et al, 15–17 April 1992.

14. I.e., 'diseases and disorders of the ears, nose, throat, and adjacent structures of the head and neck.' Anderson, 1994, p. 1133.

15. Defined in ibid., p. 657, as 'an inflammation of the lining of the stomach.'

16. Presumably what is meant by this term is damage to the function rather than the structure of the network of organs that are responsible for pumping blood throughout the body. Anderson, 1994, p. 271.

17. Asthenia signifies a lack or loss of strength and energy, and general feebleness. See ibid., p. 136.

18. Anderson, 1994, p. 646, defines functional murmur as follows: 'a heart murmur caused by an alteration of function without structural heart disease or damage, as in a murmur related to anemia rather than an organic heart disorder.'

19. Arinchin and Nalivaiko, in Nesvetailov, 1991, p. 45.

20. A neoplasm is any abnormal growth of tissue, whether benevolent or malignant.

21. Hypertonia is 'abnormally increased muscle tone or strength.' Anderson, 1994, p. 771.

22. Demichev, 1993. pp. 28–9.

23. IRM, 1992, p. 8.
24. *Katastrofa*, 24 April 1993.
25. *Svetlahorskiya naviny*, 11 November 1993.
26. *Gomel'skaya pravda*, 7 September 1993.
27. Kazakov, 18 May 1994.
28. Kanaplya, 14 April 1993.
29. 'A complex disorder of carbohydrate, fat, and protein metabolism that is primarily a result of a relative or complete lack of insulin secretion by the beta cells of the pancreas or of defects of the insulin receptors .... The onset of diabetes mellitus ... is sudden in children and usually insidious in non-insulin dependent diabetes mellitus (type II).' Anderson, 1994, p. 469.
30. Chesnov, 1993; Radyuk, 1993.
31. 'Chernobyl'skaya katastrofa,' in *SPCII*, 1994, p. 6. Another source points out that cases of diabetes among children in the republic increased from 5.1 cases per 100 000 in 1986 to 7.9 in 1992. In Homel' region, the rise was from 23.8 to 51.6. Radyuk et al., 1994.
32. Kazakov, 18 May 1994.
33. Byelookaya, 1991, p. 69. Such a phenomenon has been noted also in Ukraine. In the spring of 1993, for example, a scientific conference at the Ukrainian Ministry of Health concluded that the percentage of healthy adults and children in the contaminated areas had declined from 47 and 53 in 1987 to 28–32 percent in adults and 27–31 percent in children over the next three years. A wide variety of illnesses were cited among the non-healthy. *Rabochaya gazeta*, 12 May 1993.
34. Minsk City Council, 1994, pp. 140–1.
35. Jovanovich, 1991, pp. 11, 14.
36. Kazakov, 18 May 1994.
37. Bebenin, 19 November 1994.
38. Kazakov, 18 May 1994.
39. The contaminated villages included Brahin and Karma in the Homel' Oblast of Belarus.
40 IAC, 1991, p. 32.
41. Ibid.
42. Jovanovich, 1991, p. 15.
43. Kanaplya, 14 April 1993.
44. On the other hand, some observers choose to ignore overwhelming evidence of health problems in the contaminated areas and accept the Report at face value even today. See, for example, Ingham, 1994, pp. 7–9.
45. The conclusions may have been defined as speculative. Yet what annoyed many individuals with whom I spoke in Belarus and Ukraine about the IAC Report was that its tone sounded so authoritative. It was couched in clipped, almost abrupt language that in its very pomposity implied admonition to local doctors and specialists for their slipshod work, and used a condescending attitude to a population so fearful of the effects of radiation. I am now of the belief that such a tone was incidental and may have resulted from the need to render the Report as concise as possible.
46. Davis, 1991, p. 14.
47. See, for example, Kryzhanovskiy, 1992, p. 4.
48. Volkov, 1991, p. 62.

49. Peresypkina, 1995, p. 1. She noted, however, that the figures by no means included all those who worked in the zone in the days, months and years after the disaster.
50. *Gomel'skaya pravda*, 8 June 1993.
51. 'Chernobyl'skaya katastrofa,' *SPCII*, pp. 5–6.
52. Gurtovenko, 1995, p. 1.
53. IRM, 1992, p. 9.
54. This term was often used in Soviet terminology. Its meaning is far from clear. 'Vegeto' is an abbreviation of 'vegetative,' i.e. relating to nutrition and growth; vascular relates to a blood vessel. Dystonia has been defined above as an impairment of muscle tone. There is no such disease in Western medical terminology.
55. *Meditsinskiy vestnik*, 25 January 1993.
56. *Vecherniy Minsk*, 28 November 1994. On the ninth anniversary of Chernobyl, the Ukrainian Health Ministry raised this total to 125 000.
57. 'Chernobyl – Five Years Later,' 1991, p. 230.
58. *Sovetskaya Belorussiya*, 8 April 1995, p. 1.
59. Peresypkina, 1995, p. 1.
60. Bobneva, 1992, p. 10.
61. Anishchenko, in Nesvetailov, 1991, p. 44.
62. Arkhangelskaya, in Nesvetailov, 1991, p. 46.
63. Radyuk, in *SPCI*, April 1992. The statement is not borne out by our earlier observation that Homel' Oblast in particular had sufficient doctors by 1994. Possibly the situation improved for the better over the 1992–94 period.
64. *Respublika*, 17 March 1993.
65. IRM, 1992, pp. 8–9.
66. The following comments are based on the report of the conference in *Nastaunitskaya hazeta*, 4 August 1993.
67. Dyadichkin, April 1994.
68. See, for example, Khomskaya, et al., in Bobneva, 1992, pp. 83–4.
69. Dr. John Jagger, personal communication with the author, 27 October 1994.
70. Ivanov, 1992, p. 1.
71. Ibid.
72. *Sem'dney*, 16 January 1993. The rise in numbers has also been verified by (the then) Minister of Health V. Kazakov, who pointed out that leukemias in the period 1981–85 rose 106.5 percent from the period 1976–80, but in the 1986–90 period, the corresponding rise was 128.6 percent. Kazakov, 15–17 April 1992.
73. Ivanov, 1992, p. 3.
74. Ivanov, 3 July 1993.
75. Karkanitsa, 24 July 1993 and ff.
76. In the previous chapter, we noted that there were some 25 000 resettlers from the contaminated zone living in Minsk. It also seems likely that some of those diagnosed with leukemia would have been obliged to travel to Minsk for medical treatment, thus further inflating the totals in the capital city.
77. See also Williams, 1994, p. 556.
78. Kanaplya, 14 April 1993.
79. Ibid., 14 April 1993.
80. Likhtarev, et al., 1993, p. 594.

81. Baverstock, et al., 1992, p. 21.
82. The figure is erroneous. The most recent census of 1989 indicated that the total population of Homel' Oblast was 1 674 000, which would raise considerably the incidence of thyroid cancer among children. State Committee of the Belorussian SSR for Statistics, 1990, p. 17. Figures supplied to me by Sergey Laptev of the Minsk City Council give a 1992 population for Homel' Oblast of 1 610 900, hence the population there appears to be declining steadily, all of which provides further proof of the high incidence of children's thyroid cancer.
83. Baverstock, et al., p. 22.
84. Ibid.
85. Shigematsu and Thiesse, 22 October 1992.
86. Kazakov, et al., 1991, p. 7.
87. Anderson, 1994, p. 547, defines endemic goiter as follows: 'an enlargement of the thyroid gland caused by the intake of inadequate amounts of dietary iodine. Iodine deprivation leads to diminished production and secretion of thyroid hormone by the gland. The pituitary gland, operating on a negative feedback system, senses the deficiency and secretes increased amounts of thyroid-stimulating hormone, causing hyperplasia and hypertrophy of the thyroid gland. The goiter may grow during the winter months and shrink during the summer months when more iodine-bearing fresh vegetables are eaten.'
88. *Svetlagorskiya naviny*, 11 September 1993. Thyroiditis is the inflammation of the thyroid gland.
89. *Svetlagorskiya naviny*, 11 September 1993.
90. Belarusian Charitable Fund 'For the Children of Chernobyl,' 1992, p. 2.
91. Papillary carcinoma is the most common form of thyroid cancer. It is a 'malignant neoplasm characterized by fingerlike projections.' See Anderson, 1994, p. 1155.
92. Kazakov, et al., 3 September 1992.
93. E.P. Demidchik, Professor, Director of the Belarusian Republican Center for Cancers of the Thyroid Gland, Head of the Department of Oncology, Minsk State Medical Institute.
94. Demidchik, 1993a.
95. 'Chernobyl'skaya katastrofa,' 1994, p. 4.
96. Demidchik, 1993b.
97. See, for example, *Robitnycha hazeta*, 12 May 1993, which notes that at the time of Chernobyl, the annual number of thyroid gland cancers among children in Ukraine was four to five cases, but in 1990, 25 cases were revealed. According to Williams, 1994, p. 556, there were 276 cases in Ukraine by the Fall of 1994 and some cases had also been revealed in 'that part of the Russian Federation close to Chernobyl,' i.e. Bryansk Oblast.
98. Demidchik, 1993a.
99. Dr. John Jagger, personal correspondence with the author, 27 October 1994.
100. Williams, 1994, p. 556.
101. Demidchik, 1993a.
102. Dr. E.P. Demidchik, personal correspondence with the author.
103. Williams, 1994, p. 556.
104. Demidchik, 1993b.

105. The following statements are based on Demidchik, 1993a, and Demidchik, 1993c.
106. One US sources notes that: 'For cancer of the thyroid, the entire gland is removed, along with surrounding structures from neck to collarbone, in a radical neck dissection. Before surgery, the basic metabolism rate is lowered to normal by giving iodine and antithyroid drugs. If a tumor is present, a frozen section of the affected tissue is examined by a pathologist. If malignant cells are found, most or all of the gland is removed.' Anderson, 1994, pp. 1554–5.
107. We have omitted here the significant rise in thyroid gland cancers among children, discussed above.
108. Stozharov, 14 December 1993.
109. Ibid.
110. Dr. Ernest McCoy, a pediatrician and Professor Emeritus of Medicine at the University of Alberta, accompanied me on a visit to two hospitals in December 1993.
111. Ushakevich, 13 December 1993.
112. How many children have traveled abroad? By the spring of 1994, the Children of Chernobyl Fund alone had sent some 60 000 children to different locations. However, not all these children were from the contaminated zones and some may have travelled two or even three times. Nonetheless the total number for all organizations may be estimated to be around 500 000, which exceeds the number of children who lived in contaminated regions in 1994–5.
113. Nevertheless, it should be noted that in January 1995, Lukashenka was photographed publicly with a group of children from Mahileu and Brest oblasts at the second Children's Hospital of the city of Minsk. The visit may have reflected less the president's interest in Chernobyl than that of health generally. During parliamentary discussions in this same month, it was prognosticated that health costs would make up about one-third of the total budget. Chernobyl would be responsible for a portion of these costs, directly or indirectly. On the hospital visit, see *Sovetskaya Belorussiya*, 10 January 1995; on projected health costs, see *Sovetskaya Belorussiya*, 13 January 1995.
114. Oxford Analytica Daily Brief, 'Belarus: Economic Gloom,' 1 February 1995.
115. *Sovetskaya Belorussiya*, 13 January, 1995.

5   PERESTROIKA AND INDEPENDENCE

1. Kyiv journalist Lyubov Kovalevskaya, for example, who was employed in Pripyat prior to the Chernobyl disaster, has stated that:

   It can be contended that the International Atomic Energy Agency exercises control, and moreover that at the request of the Ukrainian and Byelorussian governments the agency's experts are investigating the contaminated areas. But we have not a few examples of post-accident investigation in the USSR – for example, at the Rovno and Crimean power stations, when the 'regiment's honour' turned out to be the main outcome, and we know that this organisation's review of the safety of plutonium produced from the 'peaceful' atom is absolutely unreliable. We know that the agency practices double bookkeeping – exactly like Soviet dosimetric control after an accident – for the benefit of the public and governments of various countries, as evidenced by the secret report of the above organisation on the review for 1987.... The above organisation conceals not merely the real problems of control, but also those of atomic power plants. Kovalevskaya, 1993, pp. 123–4.

2. Roman Yakovlevsky, in *Minsk Economic News*, May 1994, p. 5.
3. See George O. Liber, *Soviet Nationality Policy, Urban Growth, and Identity Change in the Ukrainian SSR 1923–1934*, Cambridge, 1992, pp. 49–66.
4. Henze, 1985, p. 27.
5. Urban and Zaprudnik, 1993, pp. 106–7.
6. During a visit to Minsk in October 1994, the author requested permission to work in the KGB archives. The request was not granted, but an interview was held with the chief archivist who was willing to show me several files. These pertained in part to alleged collaboration of Belarusians in Minsk with the Germans at the beginning of the war, including photographs of overtly pro-Hitler demonstrations. There is no possibility of confirming the validity of the photographs and this writer would be the first to acknowledge that photographs from the Soviet period are one of the least reliable forms of historical evidence.
7. *Zvyazda*, 14 July 1990.
8. *Moscow News*, No. 27, 2 July 1989, p. 2. See also Mihalisko, 1989.
9. *Shag*, 23 March 1990.
10. Did not the same sort of response occur elsewhere, for example in Ukraine? The answer is yes, but in Ukraine non-Communist representation in parliament was much stronger and more influential. In addition, the split within the Communist Party of Ukraine between the reformist wing in parliament (under Leonid Kravchuk) and the rank-and-file membership (under Stanislav Hurenko) was a crucial factor in the survival and operations of the Rukh (Popular Movement).
11. Adamovich, 1989, p. 12.
12. Zhuravlyov, 1989, p. 11.
13. The BPF was officially registered only on 19 June 1991. *Naviny BNF*, July 1991.
14. The feeling was, to some extent, mutual. Shushkevich remained personally hostile to Paz'nyak. He remarked in one interview that he could not bring himself to admire Paz'nyak, and that the latter had unjustly accused him of being a 'disguised apparatchik.' *Zvyazda*, 12 December 1992. Cited in *Belarusian Review*, Fall 1993, p. 2.
15. *Naviny BNF*, May 1993. A slightly different perspective is offered by Urban and Zaprudnik. They note the illegalities conducted by the authorities during the election campaign, particularly the prevention of places on the ballot for hundreds of potential opposition candidates. However, they do not perceive the elections as an unqualified success for the nomenklatura either, and note that the latter was 'decimated at the polls.' They note that only 25 per cent of party and government officials on the ballots were elected, and that the opposition was able to make significant inroads into the leadership

of the Supreme Soviet, particularly the appointment of Shushkevich as the
First Deputy Chairperson. See Urban and Zaprudnik, 1993, pp. 112–13.

16. IISEPS, 1995, p. 2.
17. *Naviny BNF*, May 1993.
18. Ibid.
19. Grushevoy, 1994, p. 5.
20. Moreover, the nuclear plant that had been the site of the disaster, and others
    of similar type, were located outside the borders of the republic. The Soviet
    authorities were thus not held responsible for 'imposing' such stations on a
    union republic, as in Ukraine or Lithuania.
21. Price, 1991, p. 4.
22. Hence the Belarusian Charitable Fund 'For the Children of Chernobyl,' the
    largest and most financially stable charitable organization in the republic,
    originated within the BPF, and its leader, Gennadiy Grushevoy, is a parlia-
    mentary deputy who ran initially on the Popular Front platform.
    Nonetheless, such an organization was developed out of necessity, not as a
    result of opposition to the government.
23. *Sovetskaya Belorussiya*, 19 January 1995.
24. Shushkevich, April 1992. In fairness to Shushkevich, it should be noted that
    he was one of the chief critics of the official secrecy surrounding Chernobyl
    and was one of the signatories of the 'Declaration of the group of experts
    working in the sphere of radiation safety and radiation medicine, on connec-
    tion with the situation arising from the accident at the Chernobyl nuclear
    power station,' of 14 September 1989. See, for example, Shushkevich,
    1990, pp. 64–5.
25. Koren' and Yaroshevich, 1993, p. 2.
26. Edchik, 1993, p. 3.
27. Koren' and Yaroshevich, 1993, p. 2 and ff.
28. Mikhalevich, 1993, p. 4 and ff.
29. For further elaboration of Mikhalevich's views, see *Minsk Economic News*,
    No. 8, April 1995, p. 4.
30. Kanaplya, 14 April 1993.
31. Stozharev, 14 December 1993.
32. *Intelnews*, 10 February 1995.
33. Stukach and Danilkin, 1993, p. 2.
34. Grushevoy, 1994, p. 5.
35. Savitsky, 1994, p. 23. See also the special issue of the journal *Demos*, 18–22
    April 1994, which features quotations from Savitsky on page one.
36. Savitsky, 1994, p. 24.
37. Rusan, 1994, pp. 26–30.
38. Drakohurst, 1995, p. 1. See also *Sem'dney*, No. 16, 22 April 1995, p. 1.
39. See, for example, *Svaboda*, 7 April 1995, p. 1. According to one source,
    Lukashenka merits a place in the *Guinness Book of Records* as the first pres-
    ident in the world unable to speak the state language of his country.
    Ogonyok, No. 15, April 1995, p. 50.
40. See, for example, *Minsk Economic News*, No. 3, February 1995, p. 4.
41. In fairness, there have been significant distortions of history on both sides,
    and the pro-Belarusian elements have often adopted a one-sided view of
    past events, such as the 1514 Battle of Orsa. Kozikis, 1995.

42. *Sovetskaya Belorussiya*, 28 March 1995, p. 1.
43. *Sem'dney*, No. 14, 8 April 1995, p. 3.
44. *Sem'dney*, No. 16, 22 April 1995, p. 3.

Bibliography

INTERVIEWS

Chesnov, Dmitriy. Chief Physician, Minsk Hospital No. 3 for Sick Children. Minsk, Belarus, 21 April 1993.

Demidchik, E.P., Professor, Director of the Belarusian Republican Center for Cancers of the Thyroid Gland, Head of the Department of Oncology, Minsk State Medical Institute. Minsk, Belarus, 17 April 1993; [1993a] and 11 December 1993. [1993c]

Glod, Vladimir, Deputy Chairman, Belinform Agency. Minsk, 9 August 1993; and 20 April 1995.

Gres', N.A., Director of the Endocrinological Center of the Institute of Radiation Medicine and Candidate of Medical Sciences. Minsk, Belarus, 14 December 1993.

Grushevaya, Irina, Chairperson of the International Association for Humanitarian Cooperation. Minsk, Belarus, 13 October 1992.

Grushevoy, Gennadiy, Chairman of the Belarusian Charitable Fund 'For the Children of Chernobyl', Deputy of the Belarusian Supreme Soviet, 13 October 1992; 20 April 1993; 11 December 1993.

Kanaplya, Evgeniy F., Professor, Director, Institute of Radiobiology, Belarusian Academy of Sciences, Deputy of the Belarusian Supreme Soviet. Minsk, Belarus, 14 April 1993.

Kavyzin, Vladimir, Chairman of the Health Commission, Minsk City Council. Minsk, Belarus, 10 December 1993.

Kazakov, V.S., Minister of Health, Republic of Belarus. Minsk, Belarus, 13 December 1993.

Kozikis, Dmitriy D., Professor of Social Studies, Minsk State Linguistic University. Minsk, Belarus, 22 April 1995.

Polishchuk, Vladimir, Chief of Minsk Hospital No. 1. Minsk, Belarus, 10 December 1993.

Radyuk, K. Physician, Minsk Hospital No. 3 for Sick Children, 21 April 1993.

Shushkevich, Stanislau. Former Chairman of the USSR Supreme Soviet. Former Pro-Rector, Minsk State University. Professor of Physics. Member of the Supreme Soviet. Minsk, Belarus, 15 April 1992.

Stozharov, A.N., Director, Institute of Radiation Medicine and Chairman, Department of Radiation Medicine and Ecology, Minsk Medical Institute. Minsk, Belarus, 14 December 1993.

Ushakevich, Iryna, Chief Physician, City of Minsk. Minsk, Belarus, 13 December 1993.

MONOGRAPHS AND ARTICLES

Academy of Sciences of the Belorussian SSR. Institute of History. *Istoriya Belorusskoy SSR*. Ed. Gorbunov, T.S. et al. Minsk, 1961.

Academy of Sciences of the Belorussian SSR. Institute of History. *Istoriya Belorusskoy SSR*. Ed. Ignatenko, I.M. et al. Minsk, 1977.

Academy of Sciences of the Belorussian SSR. Institute of History. *Istoriya rabochego klassa Belorusskoy SSR. T.4. Rabochii klass BSSR na etape sovershenstvovaniya sotsializma (1961–1986)*. Minsk, 1987.

Academy of Sciences of the Belorussian SSR. Institute of History. *Kul'turnoe stroitel'stvo v Belorusskoy SSR (1946–1958 gg.)*. Ed. Marchenko, I.E. et al. Minsk, 1979.

Adamovich, Ales'. 'Byelorussia's Calamity: Chernobyl Victims in Minsk.' *Moscow News*, No. 41, 8 October 1989, p. 12.

Afanas'ev, Anatoliy. 'Poka stroili AES, o demontazhe ikh v budushchem ne dumali.' *Inzhernerskaya gazeta*, No. 24, March 1993, p. 2.

Aleksandrov, A. 'Na strontsiy s shapkamy?' *Kommunist Belorussii*, No. 5 (1990): 81–4.

Anderson, Kenneth N., Revision Editor; Anderson, Lois E., Consulting Editor and Writer; and Walter D. Glanze, Consulting and Pronunciation Editor. *Mosby's Medical, Nursing, and Allied Health Dictionary*. Fourth Edition. St. Louis, Mo., 1994. [Cited as Anderson, 1994.]

Anishchenko, E.V. 'Public Evaluation of the Radioactive Situation in the Areas with Higher Level of Radioactive Contamination.' In Nesvetailov, 1991, p. 44.

Antonovich, Ivan. 'Natsional'nyy vopros – yavlenie mirovoe ...' *Kommunist Belorussii*, No. 3 (1990): 85–90.

Antonovych, Slavomir. *Petr Masherov*. Minsk, 1993.

Arinchin, A.N. and G.V. Nalivaiko. 'Characteristics of Bioelectric Heart Activity in Children from the Areas with Radionuclide Contamination.' In Nesvetailov, 1991, p. 45.

Arkhangelskaya, G.V., Yu.M. Lev and V.I. Usoltsev. 'The Effectiveness of Propagating Radiation Sanitary Knowledge Among the Population and Medical Personnel.' In Nesvetailov, 1991, p. 46.

Association of Byelorussians in Great Britain. *Listy da Gorbachova (Letters to Gorbachev: New Documents from Soviet Byelorussia)*. London, 1987.

Badey, G. 'Chemu otdat' prioritet?' *Kommunist Belorussii*, No. 11 (1988): 72–7.

Bahdanovic, V., U. Novik and J. Zaprudnik, 'Belarus' – New Land of Opportunity.' *Belarusian Review* (December 1992): 3.

Bandarchyk, Vasiliy Kirylavich. *Etnicheskie protsessy i obraz zhizni: Na materialakh issledovaniya naseleniya gorodov BSSR*. Minsk, 1980.

Baverstock, Keith, Bruno Egloff, Aldo Pinchera, Charles Ruchti and Dillwyn Williams. 'Thyroid Cancer After Chernobyl.' *Nature*, Vol. 359, 3 September 1992, pp. 21–2.

Bebenin, Oleg. 'Ot vzryva reaktora do sotsial'nogo vzryva ne tak uzh i daleko.' *Respublika*, 18 November 1994.

Belarusian Charitable Fund 'For the Children of Chernobyl.' 'Analysis of Morbidity on Children's Endocrinology in Belarus.' Unpublished paper, 1992.

Belarusian Popular Front 'Adrazhennie'. *Program*. Minsk, 30 May 1993.

Belarusian Sociological Service 'Public Opinion.' *Belarus': dva goda nezavisimosti (analiticheskie materialy po rezul'tatam sotsiologicheskikh issledovaniy)*. Minsk, 1992.

Belorusskaya SSR: administrativno-territorial'noe delenie. Minsk, 1968.

Belorusskaya SSR: kratkaya entsiklopediya. Ed. Brovka, Piatrus. Minsk, 1979.

Bespamyatnykh, Nikolay and Tat'yana Bogush. 'Nazovesh' li menya bratom?: Natsional'nye otnosheniya v zerkale sotsiologii.' *Kommunist Belorussii,* No. 7 (1990): 36–40.

Bespamyatnykh, Nikolay and Tat'yana Bogush. 'Ne vykidyvayte belyy flag, po sushchestvu zanovo reshaya slozhnye natsional'nye problemy.' *Kommunist Belorussii,* No. 8 (1991): 89–93.

Bobneva, M.I., ed. *Chernobyl'skiy sled: mediko-psikhologicheskie posledstviya radiatsionnogo vozdeystviya.* 2 vols. Moscow, 1992.

Borisovna-Osipova, Mariya. 'Menya ne slyshat … ' *My i vremya: Nezavisimaya levaya gazeta,* No. 19 (September 1992), p. 3.

Borodina, V.P., V.A. Dementyev, K.I. Lukashev, F.S. Martinkevich, S.M. Melnichuk, N.T. Romanovsky, A.H. Shklyar, O.F. Yakushko, and V.A. Zhuchkevich. *Soviet Byelorussia.* Moscow, 1972.

Braim, I.N. 'Sotsial'noe razvitie gorodov.' In Bandarchik, 1980, p. 31.

Brovikov, V.I., ed. *Kommunisticheskaya partiya Belorussii v rezolyutsiyakh i resheniyakh sezdov i plenumov TsK.* Minsk, 1983–4.

Byelookaya, T.V. 'Health of Byelorussian Children in the Current Ecological Situation.' In Nesvetailov, 1991, pp. 68–71.

Byelorussian SSR. Central Executive Committee. Central National Commission. *Prakticheskoe razreshenie natsional'nogo voprosa v Belorusskoy SSR.* Minsk, 1927–8.

Byelorussian SSR. Central Statistical Administration. *Belorusskaya SSR v tsifrakh v 1969 godu: statisticheskiy sbornik.* Minsk, 1970.

Byelorussian SSR. Central Statistical Administration. *Narodnoe khozyaystvo Belorussii.* Minsk, 1984.

Byelorussian SSR. Central Statistical Administration. *Narodnoe khozyaystvo Belorusskoy SSR.* Minsk, 1979–80.

Byelorussian SSR. Central Statistical Administration. *Promyshlennost' Belorusskoy SSR: statisticheskiy sbornik.* Ed. P.A. Bosyakov. Minsk, 1976.

Byelorussian SSR. Central Statistical Administration. *Razvitie narodnogo khozyaystva Belorusskoy SSR za 20 let (1944–1963gg.).* Minsk, 1964.

Chekha, Georgiy. 'Chernobyl'skiye fondy: kto i kak zanimayetsya blagotvoritel'nost'yu?' *Narodnaya hazeta,* 14 July 1993.

'Chernobyl'skaya katastrofa, ee posledstviya i puti preodoleniya na territorii Belarusi: Natsional'nyy doklad.' In *Svet paslya Charnobylya,* 1992, pp. 1–10. [*SPCII*]

Chernobyl'skaya katastrofa: prichiny i posledstviya (ekspertnoe zaklyuchenie). Chast' tret'ya: Posledstviya katastrofy na Chernobyl'skoy AES dlya Respubliki Belarus'. Ed. V.B. Nesterenko. Minsk, 1992.

'Chislennost' naseleniya BSSR.' *Kommunist Belorussii,* No. 8 (1990): 30, 34.

'Chornobyl – Five Years Later.' Editorial. *The Ukrainian Quarterly,* Vol. 47, No. 3 (Fall 1991): 229–32.

Committee of the Byelorussian SSR for Statistics. *Narodnoe khozyaystvo Belorusskoy SSR v 1989g.: statisticheskiy ezhegodnik.* Minsk, 1990.

Committee of Land Measurement with the Council of Ministers, Republic of Belarus. *Mogilevskaya Oblast'. Obzorno-topograficheskaya karta c dannymi radiatsionnogo zagryazneniya.* Minsk, 1994.

Committee of Land Measurement with the Council of Ministers, Republic of Belarus. *Respublika Belarus': ahlyadna-tapahrafichnaya karta.* Minsk, 1992.

Danilov, Aleksandr. 'Budushchee stremitsya v emigratsiyu, i eto trevozhno.' *Belaruskaya dumka*, No. 5 (1993): 58–62.

Davis, Andrew M. 'Health Care after Chernobyl: Radiation, Scarce Resources, and Fear.' Draft paper. 1991.

Dementeya, N.I. 'O 50-leti vossoedineniya Zapadnoy Belorussii c Belorusskoy SSR.' *Sovetskaya Belorussiya*, 12 November 1989, pp. 1–2.

Dementeya, N.I. and F.P. Senko, eds. *Sel's'koye khozyaystvo Belorussii.* Minsk, 1980.

Demichev, Dmitriy. 'Molokh sam ne ostanovitsya: Mestnye Sovety v preodolenii posledstviy Chernobyl'skoy katastrofy.' *Belaruskaya dumka*, No. 12 (1993): 27–9.

Demidchik, E.P. 'Osobennosti klinicheskogo techeniya i khirurgicheskogo lecheniya raka shchitovidnoy zhelezy u detey.' Unpublished paper. Minsk, September 1993. [1993b]

Demidchik, E.P. 'Thyroid Cancer in Belarus, 1985–1994.' Unpublished statistical data. Minsk, 1995.

Drakohurst, Yury. 'Opposition Left Out in the Cold: Totalitarianism Strikes Hunger-Strikers.' *Minsk Economic News*, No. 8, April 1995, pp. 1, 3.

Drits, V.I., ed. *Sotsial'naya infrastruktura i sotsialisticheskiy obraz zhizni.* Minsk, 1983.

Duss, Sergei. 'NGOs and Belarusian Charitable Organizations.' Draft of paper presented at the First International Congress, 'The World After Chernobyl.' Minsk, April 1992.

Dyadichkin, V. 'Functional Health Screening of Secondary School Students During the Pre- and Post-Chernobyl Periods.' Draft of a paper delivered at the Second International Congress, 'The World After Chernobyl.' Minsk, April 1994.

Edchik, I. 'Nuzhny li Belarusi AES?' *Femida*, No. 17, 26 April–2 May 1993, p. 2.

Eshin, S.Z. *Razvitie kul'tury v BSSR za gody Sovetskoy vlasti.* Minsk, 1970.

Eventov, Mark Arkadevich. *60 geroicheskikh let.* Minsk, 1978.

Gilitskiy, F.I. *Sotsial'no-ekonomicheskie problemy trudovykh resursov Belorusskoy SSR.* Minsk, 1977.

Gol'bin, Ya.A., ed. *Trudovoy potentsial Belorusskoy SSR v usloviyakh intensifikatsiy.* Minsk, 1988.

Golenchenko, G.Ya. and V.P. Osmolovskiy. *Istoriya Belarusi: Voprosy i otvety.* Minsk, 1993.

'Gosudarstvennaya programma likvidatsii v Belorusskoy SSR posledstviy avarii na Chernobyl'skoy AES na 1990–1995 gody. ' *Sovetskaya Belorussiya*, 21 October 1989, pp. 1–2.

Gosudarstvenniy doklad: o sostoyaniy okruzhayushchey prirodnoy sredy Rossiiskoy federatsii v 1991 godu. Moscow, 1992.

Grakhovskiy, A.A. 'Y cherty otchuzhdeniya.' *Kommunist Belorussii*, No. 5 (1990): 74–7.

Grishan, Igor. 'Kuropaty: sledstvie vozobnovleno: [Delo o rasstrelakh grazhdan v Kuropatakh].' *Sovetskaya Belorussiya*, 20 March 1993.

Grushevoy, Gennadiy. 'Chernobyl is still a bitter reality.' *Minsk Economic News*, May 1994: 5.

Gurtovenko, Yuriy. 'Naedine s ChP, kotoromu ne vidno kontsa.' *Sovetskaya Belorussiya*, 28 March 1995, p. 1.
Guthier, Steven L. 'The Belorussians: National Identification and Assimilation, 1897–1970.' Part 1: *Soviet Studies*, Vol. XXIX, No. 1 (January 1977): 37–61. [1977a] Part 2: *Soviet Studies*, Vol. XXIX, No. 2 (April 1977): 270–83. [1977b]
Haranski, M.M. *Ekanomika Belarusi uchora i sionnia.* Minsk, 1972.
Henze, Paul B. 'The Spectre and Implications of Internal Nationalist Dissent: Historical and Functional Comparisons.' In S. Enders Wimbush, ed., *Soviet Nationalities in Strategic Perspective.* London, 1985.
Homeyer, Burkhard. 'Internationale humanitaere Zusammenarbeit – Einsichten under Ergebnisse.' *SPCII*, pp. 98–103.
Ignatouski, V.M. *Karotki narys histor'ii Belarusi.* Minsk, 1992.
Independent Institute of Socio-Economic and Political Studies. 'Belarusian Popular Front: Rise of a Political Party.' *Minsk Economic News*, No. 2, January 1995, p. 2. [IISEPS, 1995]
Ingham, Bernard. 'Back to the future.' *Nuclear Forum* (Autumn 1994): 7–9.
Institute of Art, Ethnography and Folklore, Academy of Sciences, BSSR. *Cultural Policy in the Byelorussian Soviet Socialist Republic.* Paris, 1979.
Institute of Radiation Medicine, Ministry of Health, Republic of Belarus. *Belarus': zhytstsye u promnyakh Charnobylya.* Minsk, 1992. [IRM]
International Advisory Committee. *The International Chernobyl Project. An Overview. Assessment of Radiological Consequences and Evaluation of Protective Measures.* International Atomic Energy Agency. Vienna, 1991. [IAC]
International Monetary Fund. *Belarus.* Washington, D.C., 1992.
Istoriya Velikoy Otechestvennoy voiny Sovetskogo Soyuza 1941–1945. Chairman of editorial board: Pospelov, P.N.
   Vol. 1, *Podgotovka i razvyazyvanie voiny imperialisticheskimi derzhavami.* Moscow, 1960.
   Vol. 2, *Otrazhenie Sovetskim narodom verolomnogo napadeniya fashistskoy Germanii na SSSR. Sozdanie usloviy dlya korennogo pereloma v voyne (iyun' 1941g.-noyabr' 1942g.).* Moscow, 1963.
   Vol. 3, *Korennoy perelom v khode Velikoy Otechestvennoy Voyny (noyabr' 1942g.-dekabr' 1943g.).* Moscow, 1961.
   Vol. 4, *Izgnanie vraga iz predelov Sovetskogo Soyuza i nachalo osvobozhdeniya narodov Evropy ot fashistskogo iga (1944 god).* Moscow, 1962.
   Vol. 5, *Pobedonosnoe okonchanie voyny s fashistskoy Germaniey. Porazhenie imperialisticheskoy Yaponii.* Moscow, 1963.
Ivanov, E.P. 'Zabolevaniya krovi posle Chernobylya.' Draft of paper presented at the First International Conference 'The World After Chernobyl.' Minsk, Belarus, April 1992.
Ivanov, Evgeniy. 'Soveshchanie Belorusskikh Gematologov: Rosta detskykh leykemiy net i v 1992 godu.' *Sem' dney*, 3 July 1993.
Iwanow, Mikolaj. 'The Byelorussians of Eastern Poland under Soviet Occupation, 1939–1941.' In Keith Sword, ed. *The Soviet Takeover of the Polish Eastern Provinces, 1939–41.* London, Macmillan, 1991, pp. 253–67.
Jovanovich, Jovan V. 'The Chernobyl Accident: Five Years After. Part II: Radioactive Releases and Consequences.' Draft paper. Winnipeg, 1991.

Kadatskiy, V.V., V.A. Kuznetsov and O.V. Kadatskaya. 'Landscape-Geochemical Concept of Studying the Chernobyl Contamination over the Byelorussian Territory.' In Nesvetailov, 1991, p. 27.

Kanoplya [Kanaplya], E.F. 'Global Ecological Consequences of the Chernobyl Nuclear Explosion.' *SPCI*, April 1992, pp. 6–15.

Kapylovich, Mikola. 'Samasely: zyamlya pad chornymi krylami.' *Respublika*, 21 April 1995, p. 5.

Karkanitsa, Leonid. 'Poka net povoda dlya optimizma.' *Sem' dney*, 24 July 1993.

Kartel, N.A. 'Biological and Radioecological Aspects of the Chernobyl Accident Consequences (Analytical Review).' In Nesvetailov, 1991, pp. 12–18.

Kasperovich, G.I. *Migratsiya naseleniya v goroda i etnicheskie protsessy*. Minsk, 1985.

Kazakov, V.S. 'My na pravil'nom puti.' *Gomel'skaya pravda*, 18 May 1994.

Kazakov, V., V. Matyuchin, L. Astakhova, E. Demidchik, E. A. Korotkevich, and E. Ivanov. 'Health Assessment of Belorussian Population Exposed to Radiation due to the Chernobyl Accident.' Unpublished proceedings; Third Republican Conference on the Evaluation of Radiation Health Consequences, Homel', Belarus, April 15–17, 1992, pp. 83–7.

Kazakov, V.S., V.A. Matyukhin, L.N. Astakhova, E.P. Demidchik, G.I. Lazyuk, E.P. Ivanov, E.A. Korotkevich, Y.Ye. Kenigsberg, V.G. Rusetskaya, Yu.I. Gavrilin, V.T. Khrushch, V.F. Minenko, N.A. Gres', V.M. Drozd, A.N. Arinchin, and S.I. Tochitskaya. 'Sostoyanie zdorov'ya naseleniya Respubliki Belarus', podvergshegosya vozdeystviyu radionuklidov v svyazi s avariey na ChAES.' In *Katastrofa na Chernobyl'skoy AES i otsenka sostoyaniya zdorov'ya naseleniya Respubliki Belarus'*. Minsk, 1991.

Kazakov, Vasili S., Evgeni P. Demidchik, and Larisa N. Astakhova. 'Thyroid Cancer after Chernobyl.' *Nature*, Vol. 359, 3 September 1992, p. 21.

Kazakov, V.S., N.A. Krysenko, A.N. Stozharov, Ya. Ye. Keningsberg, L.N. Astakhova and S.N. Nalivko. 'Osnovnye rezul'taty nauchnykh issledovaniy poluchennye pri vypolnenii meditsinskogo razdela gosprogrammy preodoleniya posledstviy katastrofy na ChAES.' *SPCII*, pp. 14–18.

Kharevskiy, Vasiliy. 'Goroda navyrost: kak sovmestit' interesy predpriyatiy i lyudey?' *Kommunist Belorussii*, No. 1 (1990): 42–5.

Khomskaya, E.D., E.V. Enikolopova, N.G. Manelis, I.S. Gorina, E.V. Budyka, and Yu.V. Malova, 'Neiropsikhologicheskiy analiz posledstvii oblucheniya mozga posle Chernobyl'skoy avarii.' In Bobneva, 1992, pp. 83–104.

Kipel, Vitaut and Zora Kipel, eds. *Byelorussian Statehood: Reader and Bibliography*. New York: Belorussian Academy of Arts and Sciences, 1988.

Kiselev, K.V., ed. *Belorusskaya SSR na mezhdunarodnoy arene*. Moscow, 1964.

Kiselev, T.Ya. *Sovetskaya Belorussiya*. Minsk, 1982.

Kommunisticheskaya partiya Belorussii v rezolyutsyakh i resheniya 1918–1970. Minsk, 1973.

Koren', G. and A. Yaroshevich. 'Al'ternativy atomu net.' *Inzhernerskaya gazeta*, No. 1, January 1993, p. 2.

Kosovich, A.'Dom v kotorom nam zhit': razdum'ya ob ekologicheskikh posledstviyakh khozyaystvovaniya.' *Kommunist Belorussii*, No. 11 (1988): 64–9.

Kostyuk, M. 'Sledovat' pravde istorii.' *Kommunist Belorussii*, No. 1 (1988): 36–41.

Kovalevskaya, Lyubov. 'Reflections on the catastrophe.' In *Beyond Chernobyl: Women Respond*. Compiled by Corin Fairburn Bass and Janet Kenny. Sydney, 1993.

Kozlovskaya, L.A. 'On Branch and Territorial Structural Transformation Problems in [*sic*] Byelorussian Economy in Connection with the Chernobyl Accident Consequences.' In Nesvetailov, 1991.

'KPB v zerkale statistiki.' *Kommunist Belorussii*, No. 5 (1990): 27–9.

Kravchenko, I.S. and I.E. Marchenko. *Radyans'ka Bilorusiya: korotkiy istoryko-ekonomichniy narys*. Kyiv, 1965.

Kryzhanovskiy, A. 'Chelovek na 'bol'noy' zemle.' *Kommunist Belorussii*, No. 2 (1989): 68–75.

Kryzhanovskiy, A. 'Mir uznal polupravdu a rasplachivats'ya budut sovsem ne te, kto zakazyval 'muzyku' ... ' *Ekho Chernobylya*, No. 5–6 (February 1992): 4.

Kryzhanovskiy, Aleksandr. 'Polyusa glasnosti.' *Kommunist Belorussii*, No. 1 (1990): 49–53.

Kuznetsov, Igor. 'Vileyskaya tragediya.' *Narodnaya hazeta*, April 8–10, 1995, p. 2.

Likhtarev, I.A., N.K. Shandala, G.M. Gulko, I.A. Kairo, and N.I. Chepurny. 'Ukrainian Thyroid Doses After the Chernobyl Accident.' *Health Physics* 64 (June 1993): 594–9.

Lubachko, Ivan S. *Belorussia under Soviet Rule 1917–1957*. Lexington, KY: University Press of Kentucky, 1972.

Lukashev, S.A. *Ideologicheskaya rabota Kompartii Belorussii v poslevoennyy period razvitiya sotsialisticheskogo obshchestva (1946–1958)*. Minsk, 1979.

Lupach, Dmitriy. 'Spasti sebya my mozhem tol'ko sami ... ' [Interview with Gennadiy Grushevoy.] *Neman*, No. 12, 1992, pp. 110–19.

Lyutsko, A.M. *Fon Chernobylya*. Minsk, 1990.

Lyutsko, A.V. *Deyatel'nost' KPB po podboru vospitaniyu i rasstanovke rukovodyashchikh kadrov (1943–1945gg.)*. Minsk, 1973.

Lyutsko, A.V. *Deyatel'nost' KPB po podboru vospitaniyu i rasstanovke rukovodyashchikh kadrov (1946–1950gg.)*. Minsk, 1975.

Malashko, A. 'V sem'e edinoy.' *Kommunist Belorussii*, No. 3 (1988): 62–9.

Mal'dis, A. 'Davayte nakonets', zagovorim ... .' *Kommunist Belorussii*, No. 3 (1989): 73–6.

Malinin, S.N. and K.I. Shabun, eds. *Ekonomicheskaya istoriya BSSR*. Minsk, 1969.

Marchenko, I.E. and V.I. Novitskiy. *Belorusskaya SSR: kursom perestroyki*. Minsk, 1989.

Marples, David R. 'Environment, Economy, and Public Health Problems in Belarus.' *Post-Soviet Geography*, Vol. 35, No. 2 (1994): 102–12. [1994a]

Marples, David R. 'Kuropaty: The Investigation of a Stalinist Historical Controversy.' *Slavic Review*, Vol. 53, No. 2 (Summer 1994): 513–23. [1994b]

Marples, David R. 'Post-Soviet Belarus' and the Impact of Chernobyl',' *Post-Soviet Geography*, Vol. XXXIII (September 1992): 419–31.

Marples, David R. *The Social Impact of the Chernobyl Disaster*. London: Macmillan, 1988.

Marples, David R. *Ukraine Under Perestroika: Ecology, Economics, and the Workers' Revolt*. London: Macmillan, 1991.

Masherov, Petr Mironovich. *Izbrannye rechi i stati*. Minsk, 1982.

Mats'ko, V.I. 'Chernobyl' ne poterpit 'dvoechnikov.' *Belarus'ka dumka*, No. 5 (1993): 55–7.

Matusevich, Anton Vladimirovich. *Mestnye Sovety: preobrazovanie dereven.* Minsk, 1983.

Mavrishchev, V.S. *Khozyaystvennyy mekhanism na putyakh perestroiki.* Minsk, 1989.

McCauley, Martin. *The Soviet Union 1917–1991.* New York, 1993.

Medvedev, Grigoriy. 'Atomnyy dzhin, ili O tom, kak kontseptsiya 'absolyutnoy bezopasnosti AES' obernulas' Chernobyl'skoy katastrofoy.' *Kommunist Belorussii*, No. 7 (1990): 55–60.

Mienski, J. 'The Establishment of the Belorussian SSR.' In Kipel and Kipel, 1988, pp. 137–77.

Mihalisko, Kathleen. 'Belorussian Popular Front Holds Founding Congress in Vilnius.' *Radio Liberty Research Bulletin*, RL 308/89, 27 June 1989.

Mikhalevich, Aleksandr. 'Atomnym elektrostantsiyam al'ternativy net.' *Respublika*, 6 April 1993.

Mikhnevich, Arnol'd Efimovich. *Russkiy yazyk v Belorussii.* Minsk, 1985.

Ministry of Higher, Middle Special and Professional Education BSSR. Belorussian State University 'V.I. Lenin.' *Sbornik statey po politekonomii.* Minsk, 1959.

Minsk City Council, Department of Health; Institute of Radiation Medicine, Ministry of Health of the Republic of Belarus; Endoecological Center. *Mediko-ekologicheskiy monitoring zdorov'ya detey i podrostkov goroda Minska.* Scientific leader N.A. Gres'. Minsk, October 1992–May 1993.

Minsk City Council, Department of Health; Institute of Radiation Medicine, Ministry of Health of the Republic of Belarus; 'Endoecological Center'. *Ekologiya i sostoyanie zdorov'ya detey goroda Minska posle katastrofy na Chernobyl'skoy atomnoy elektrostantsii. Mediko-ekologicheskiy monitoring – 1994.* Scientific leader N.A. Gres'. Minsk, 1994.

Myalitski, Mikola. 'Nas pachuyuts' u patsyarpelykh raenakh.' *Litaratura i mastatstva*, 8 October 1993.

Myasnikov, Anatoliy. 'Kuropaty raskryvayut tayny.' *Kommunist Belorussii*, No. 8 (1990): 92–3.

Myasnikovich, Mikhail. 'Ekonomicheskiy soyuz–novaya forma ekonomicheskoy integratsii.' *Belarus'kaya dumka*, No. 5 (1993): 1–8.

'Natsional'nyy sostav naselennya g. Minska.' *Kommunist Belorussii*, No. 9 (1988): 44.

Nesvetailov, G.A., editor-in-chief. Children of Chernobyl Byelorussia Committee. *Chernobyl Digest of Current Relevant Literature.* Moscow-Minsk, 1991.

Olekhnovich, G.I. *Ekonomika Belorussii v usloviyakh Velikoy Otechestvennoy Voiny (1941–1945).* Minsk, 1982.

Padluzhny, Alyaksandr. 'Kab rodnym slovam smahu natalits': mounaya palitika na suchasnym etape.' *Belaruskaya dumka*, No. 5 (1993): 77–84.

Paz'nyak, Zyanon. 'Dvuyazychie i byurokratizm.' *Raduga*, No. 4, 1988.

Paz'nyak, Zyanon. *Sapraudnae ablichcha.* Minsk, 1992.

Paz'nyak, Z. and Ya. Shmyhaleu. 'Kurapaty: daroga smertsi.' *Litaratura i mastatstva*, 3 June 1988.

Peresypkina, Lidiya. 'Chernobyl'skaya programma ne budet svernuta.' *Sovetskaya Belorussiya*, 26 April 1995, p. 1.

Petrayev, E. and S.L. Leynova. 'Active Particles is [*sic*] a Typical Feature of Contamination in the South Regions of Byelorussia.' In Nesvetailov, 1991, p. 21.

Petrayev, Evgeniy. 'Radiatsionnaya obstanovka menyaetsya v luchshuyu storonu, no rasslabyat'sya nel'zya.' *Holas Vetkaushchyny*, No. 23, 1993.

Petrikov, Petr Tikhonovich. *Dostizheniya istoricheskoy nauki v BSSR za 60 let.* Minsk, 1979.

Pikulik, M.M. 'The Effects of Chernobyl on the Animal Life of Belarus.' Draft of paper presented at the First International Congress, 'The World After Chernobyl.' Minsk, April 1992.

Pokshishevsky, V.V. *Geography of the Soviet Union.* Moscow, 1974.

Price, Joe. 'Byelorussian Independence: Thoughts and Observations.' *Byelorussian Review* (Summer 1991): 3–7.

Prognoz Center. Belorusskiy tsentr sotsial'nogo i politcheskogo prognoza. *Analiticheskiy Byulleten': Vypusk 1, Yanvar'-Fevral' 1993 goda.* Minsk, 1993.

Radyuk, K.A. 'The Psychological Effects of Chernobyl.' Draft of paper presented at the First International Congress, 'The World After Chernobyl.' Minsk, April 1992.

Radyuk, K., L. Bortkevich, and I. Kunavich. 'Status of Health of Children Suffering from Diabetes Mellitus after the Chernobyl Disaster.' Draft of paper presented at the Second International Congress, 'The World After Chernobyl.' Minsk, April 1994.

Rosenberg, William G. and Lewis H. Siegelbaum, eds. *Social Dimensions of Soviet Industrialization.* Bloomington, Ind., 1993.

Rusan, V.I. 'Perspektivy ekologicheski chistoy energetiki i energoobespecheniya APK.' In *SPCII*, 18–22 April 1994, pp. 26–30.

Rutskaya, E. 'Rodnaya rech' – eto bestsenyy klad.' *Kommunist Belorussii*, No. 3 (1989): 77–8.

Savastenko, Aleksandr. 'K garmonii ili k katastrofe?' *Belaruskaya dumka*, No. 2 (February 1993): 50–4.

Savitsky, B.P. 'Belarus' i atomnaya energiya: Problemy i perspektivy.' In *SPCII*, pp. 22–5.

Semkin, Aleksandr. 'Osobaya zona – bez osobogo vnimaniya.' *Belaruskaya dumka*, No. 1 (1994): 21–4.

Shabailov, V.I., et al. *Belorusskaya SSR: status, dostizheniya, razvitie.* Minsk, 1989.

Shabel, S., Danilov, A., Kuz'muk, A. and O. Kravchenko. 'Molodezh' i rynok.' *Kommunist Belorussii*, No. 7 (1991): 28–37.

Shakhotko, L.P. *Rozhdaemost v Belorussii: sotsial'no-ekonomicheskie voprosy.* Minsk, 1975.

Shigematsu, I, and J.W. Thiessen. 'Childhood Thyroid Cancer in Belarus.' *Nature*, Vol. 359, 22 October 1992, p. 681.

Shimov, Vladimir and Yakub Aleksandrovich. 'Traektoriya vo vremeni: perspektivy ekonomiki Belarusi.' *Belaruskaya dumka*, No. 4 (1993): 13–21.

Shumarova, E. 'Bragin: Relocation by Lot.' In Nesvetailov, 1991, p. 64.

Shushkevich, S.S. 'Vremya poluraspada tayn: Chernobyl': problemy i pokhody k izh razresheniyu.' *Kommunist Belorussii*, No. 2 (1990): 64–8.

Shytkov, G.V., et al. *Istoriya Belorusskoy SSR.* Minsk, 1977.

Sidorenko, Olexandr. 'The Chornobyl Tragedy: Its Consequences as Estimated by Japanese Scientists.' *Ukraine*, No. 4 (1989): 40–1.

Smirnov, Yu.P. and R.P. Platonov. *Na krutom perelome: Ocherk osnovnykh napravlenii deyatelnosti Kompartii Belorussii po osushchestvleniyu perestroyki*. Minsk, 1989.

Smolenskiy, A. 'Legko byt' illyuzionistom.' *Kommunist Belorussii*, No. 5 (1989): 15–16.

Smolich, Arkadz'. *Heohrafiya Belarusi*. Minsk, 1993.

Solchanyk, Roman. 'Belorussia and Belorussians in the USSR: Nationality and Language Aspects of the Census of 1979.' *Radio Liberty Research Bulletin*, RL 115/80, 21 March 1980. [1980b]

Solchanyk, Roman. 'Belorussia: The State of the Republic.' *Radio Liberty Research Bulletin*, RL 218/78, 8 October 1978.

Solchanyk, Roman. 'A Belorussian Scholar on the Role of the Native Language.' *Radio Liberty Research Bulletin*, RL 222/79, 24 July 1979.

Solchanyk, Roman. 'The Study of the Russian Language in Belorussia.' *Radio Liberty Research Bulletin*, RL 30/80, 21 January 1980. [1980a]

State Committee of the Belorussian SSR for Statistics. *Narodnoe khozyaystvo Belorusskoy SSR v 1989 g*. Minsk, 1990.

Stukach, Vladimir and Viktor Danilkin. 'Tenisarkofaga, ili Novye strasti vokrug atomnoy energetiki.' *Femida*, No. 17, 26 April–2 May 1993, p. 2.

'Svet paslya Charnolbyla.' International Congress: The World after Chernobyl. Congress Section on Medical Scientific Reports. Minsk, April 16–20, 1992. [*SPCI*]

'Svet paslya Charnobylya.' II Mizhnarodny Kanhres. *Nauchnye tezisy k Kongressu*. [The Second International Congress: 'The World after Chernobyl' Scientific Reports.] Minsk, 18–22 April 1994. [*SPCII*]

Taranovskiy, G.S. *Kuropaty: sledstvie prodolzhaetsia*. Moscow, 1990.

Trusov, O. 'Pochemu tol'ko 'bilingvizm'?' *Kommunist Belorussii*, No. 5 (1989): 15.

Turevich, Art. 'Belarus to Sign the Collective Security Treaty.' *Belarusian Review* (Spring 1993): 5.

Udovenko, S.A. 'Situation and Prospects for Protection of the BSSR Territory from Radiocontamination.' In Nesvetailov, 1991, p. 20.

Urban, Michael E. *An Algebra of Soviet Power: Elite Circulation in the Belorussian Republic, 1966–86*. Cambridge, UK, 1989.

Urban, Michael and Jan Zaprudnik. 'Belarus: A Long Road to Nationhood.' In Bremmer, Ian and Ray Taras, eds. *Nations and Politics in the Soviet Successor States*. Cambridge, UK, 1993.

Urban, Paul. 'Byelorussian Political Activities and the Establishment of the Byelorussian Soviet Socialist Republic.' In Kipel and Kipel, 1988, pp. 179–203.

Vakar, Nicholas. *Belorussia: The Making of a Nation*. Cambridge, Mass., 1956.

Vakar. Nicholas P. 'The Belorussian People between Nationhood and Extinction.' In Goldhagen, Erich, ed. *Ethnic Minorities in the Soviet Union*. New York, 1968.

Vardys, V. Stanley. 'The Baltic States under Stalin: The First Experiences, 1940–41.' In Sword, Keith, ed. *The Soviet Takeover of the Polish Eastern Provinces, 1939–41*. London: Macmillan, 1991, pp. 268–90.

Velikhov, E. 'Budem pokupat' AES za rubezhom?' *Inzhernerskaya gazeta*, No. 80, 1992.

Volkov, A.E. 'Katastrofa, kotory net kontsa.' *Kommunist Belorussii*, No. 8 (1991): 60–5.

Volodin, O. 'Predmet vseobshchei zaboty i vnimaniya.' *Narodnoe obrazovanie*, No. 9 (1979): 39–48.

Vyachorka, V.R. *Pra herb i stysah*. Minsk, 1993.

Williams, Dillwyn. 'Chernobyl, Eight Years On.' *Nature*, Vol. 371, 13. October 1994, p. 556.

Yachenko, S.E., et al. *Nekotorye voprosy ekonomicheskogo razvitiya Belorussii*. Minsk, 1969.

Yakovlevsky, Roman. 'Faces and Images.' *Minsk Economic News*, April 1993, p. 5.

Yakushev, B.I. 'Radioactive Contamination of Plant Life.' Draft of paper presented at the First International Congress, 'The World after Chernobyl.' Minsk, April 1992.

Yakutov, Vladimir. *Petr Masherov: khudozhestvenno-dokumental'naya povest'*. Minsk, 1992.

Yanovich, Sakrat. 'Lukashenka: chalavek 'bez natsyyanal'nas'tsi'.' *Svaboda*, 24 March 1995, p. 2.

Yaros', Anatoliy. 'Agoniya prodolzhaetsya.' *Kommunist Belorussii*, No. 5 (1980): 78–80.

Zagorodnyuk, I. Kh. 'Kuropaty: fal'sifikatsiya veka?' *My i vremya: Nezavisimaya levaya gazeta*, No. 19, September 1992, p. 2.

Zaprudnik, Jan. *Belarus: At a Crossroads in History*. Boulder, Colo.: Westview, 1993.

Zhilinskiy, N. 'S Leninskoy mysl'yu.' *Kommunist Belorussii*, No. 4, 1990, pp. 17–21.

Zhukovskiy, V. and V. Talapin. ''Beskhoznaya' ekologiya, ili nekotoraya informatsiya dlya vozmushcheniya.' *Kommunist Belorussii*, No. 6, 1989, pp. 77–81.

Zhuravlyov, Alexander. 'Not to Miss a Chance. Byelorussia: Prospects of Development.' *Moscow News*, No. 48, 26 November 1989, p. 11.

Zverev, L. and G. Grushevoy. 'Budet li vypolnena natsional'naya geneticheskaya programma?' *Narodnaya hazeta*, 16 December 1992.

NEWSPAPERS AND JOURNALS

Belarus News (Minsk)
Belarusian Review (originally *Byelorussian Review*, Los Angeles, USA)
Belaruskaya dumka (Minsk)
Belorusskiy rynok (Minsk)
Chacherski vesnik (Chachersk, Homel' Oblast)
Demos (Minsk)
Dobryy vecher [*vechar*] (Minsk)
Ekho Chernobylya (Kyiv)
Ekologiya Minska
Femida (Minsk)

Golos Krasnopol'shchyny (Krasnapolle, Mahileu Oblast)
Gomel'skaya pravda (Homel')
Grodnenskaya pravda (Hrodna)
Holas Vetkaushchyny (Vetka, Homel' Oblast)
Intelnews (Kyiv)
Inzhernerskaya gazeta (Moscow)
Kommunist Belorussii (Minsk–see also *Belaruskaya dumka*)
Mahileuskaya prauda (Mahileu)
Meditsinskaya gazeta (Moscow)
Meditsinskiy vestnik (Minsk)
Minsk Economic News
Mogilevskie vedomosti (Mahileu)
Moscow News
Narodnaya hazeta (Minsk)
Nastaunitskaya hazeta (Minsk)
Naviny BNF (Minsk)
Neman (Minsk)
Ogonek (Moscow)
Oxford Analytica (UK)
Pravda (Moscow)
Raduga (Tallinn)
Rabochaya gazeta (Kyiv)
Respublika (Minsk)
Sem'dney (Minsk)
Shag (Minsk)
Sovetskaya Belorussiya (Minsk)
Svetlahorskiya naviny (Svetlahorsk, Homel' Oblast)
Vecherniy Minsk (Minsk)
Vitebskiy rabochiy (Vitebsk)
Zvyazda (Minsk)

Glossary

apparatchik	one who works in the government or party bureaucracy
BPF	Belarus Popular Front
CPB	Communist Party of Belarus (Belorussia)
CPSU	Communist Party of the Soviet Union
dacha	country house or cottage
gubernia	province (pre-Soviet period)
kulak	rich peasant
KGB	Committee for State Security
MTS	Machine Tractor Station
NKVD	People's Commissariat for Internal Affairs
nomenklatura	Communist Party ruling elite
oblast	province
Ostarbeiter	(lit.) Eastern workers; those sent to Germany as forced labor for the Reich in the Second World War
raion	district
samosely	villagers who returned to evacuated regions without the permission of the authorities
subbotnik	voluntary Saturday worker

Index

Belarus, one of the least known of the newly-independent states of the former Soviet Union, received the bulk of the radioactive fallout from the 1986 accident at Chernobyl. This book is based on extensive research work during seven visits to Belarus between 1992 and 1995. It seeks to address two major questions: has Soviet rule in Belarus retarded national cultural and state development to the extent that the republic is incapable of dealing with a major disaster; and second, what have been the main medical and social repercussions of that event? In answering these questions, David Marples provides a detailed perspective of a republic in a deep crisis and struggling for national political survival.

As the first major study on the health consequences of Chernobyl, ten years after that event, it highlights the impact of radioactive fallout on clean-up workers, evacuees and those who today live in the contaminated zones. It illustrates that the main burden of a nuclear disaster is being borne by those least able to cope with its consequences and at a time when a new country appears ready to self-destruct.

———————

David R. Marples is Professor of Russian History at the University of Alberta, Canada. He is the author of four books and numerous articles on Soviet and post-Soviet affairs, including *Ukraine under Perestroika* and *Stalinism in Ukraine in the 1940s*.

———————

The cover-design incorporates a map of Belarus and the surrounding countries (by kind permission of Joe Arciuch, *Belarussian Review*).

ISBN 0-333-62632-X

9 780333 626320

90101